D0898116

JAPANESE AND WESTERN BIOETHICS

Philosophy and Medicine

VOLUME 54

The titles published in this series are listed at the end of this volume.

JAPANESE AND WESTERN BIOETHICS

STUDIES IN MORAL DIVERSITY

Edited by

KAZUMASA HOSHINO

KLUWER ACADEMIC PUBLISHERS

DORDRECHT / BOSTON / LONDON

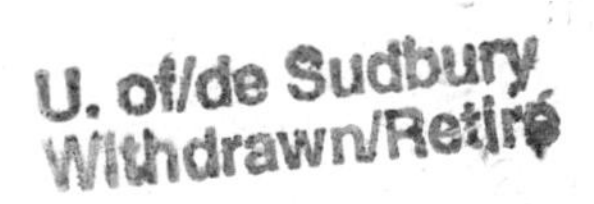

Library of Congress Cataloging-in-Publication Data

Japanese and Western bioethics : studies in moral diversity / edited by Kazumasa Hoshino.
p. cm. -- (Philosophy and medicine ; v. 54)
Proceedings of the United States-Japan Bioethics Congress held Sept. 2-4, 1994, in Tokyo, Japan.
Includes index.
ISBN 0-7923-4112-0
1. Medical ethics--Japan--Congresses. 2. Bioethics--Japan--Congresses. 3. Medical ethics--Cross-cultural studies--Congresses. 4. Bioethics--Cross-cultural studies--Congresses. 5. Ethics, Comparative--Congresses. I. Hoshino, Kazumasa. II. United States-Japan Bioethics Congress (1994 : Tokyo, Japan) III. Series.
R724J36 1996
174'.2'0952--dc20 96-22746

ISBN 0-7923-4112-0

Published by Kluwer Academic Publishers,
P.O. Box 17, 3300 AA Dordrecht, The Netherlands

Kluwer Academic Publishers incorporates
the publishing programmes of
D. Reidel, Martinus Nijhoff, Dr W. Junk and MTP Press.

Sold and distributed in the U.S.A. and Canada
by Kluwer Academic Publishers,
101 Philip Drive, Norwell, MA 02061, U.S.A.

In all other countries, sold and distributed
by Kluwer Academic Publishers Group,
P.O. Box 322, 3300 AH Dordrecht, The Netherlands

Printed on acid-free paper

Printed in the Netherlands

CONTENTS

FOREWORD

The editors of the Philosophy and Medicine series recognize with gratitude the foresight, understanding, hard labor, and patience of Prof. Kazumasa Hoshino. It is his perseverance that has made this volume a reality. It was his faith in ideas that brought together a cluster of scholars in Tokyo on September 2–4, 1994, at Sophia University for a U.S.–Japan Bioethics Congress. With the support of the Foundation for Advancement of International Science, the Japan Foundation Center for Global Partnership, the Foundation of Thanatology, the Japanese Center for Quality of Life Studies, and Sophia University, scholars from Canada, Germany, Japan, and the United States were able to explore the differences and similarities in their approaches to bioethics and health care policy. That conference first produced a volume through Shibunkaku Publishers of Kyoto that appeared in 1995 in Japanese: *The Dignity of Death*, edited by Kazumasa Hoshino. Selections from those materials have been reworked for an English audience and now appear, along with new essays, in this volume. The field of comparative bioethics is only in its infancy. We are deeply grateful to Prof. Kazumasa Hoshino, one of the fathers of Japanese bioethics, for having made this volume possible.

H. Tristram Engelhardt, Jr.
Stuart F. Spicker

ACKNOWLEDGEMENTS

This volume's editors and Kluwer Academic Publishers wish to thank Shibunkaku Press, Kyoto, Japan, for permission to publish, without charge, essays derived from the U.S.–Japan Bioethics Congress held in Tokyo, Japan, September 2–4, 1994. Shibunkaku Press published these as a proceedings volume: *The Dignity of Death* (ed. Kazumasa Hoshino, Kyoto: Shibunkaku Press, 1995). We are grateful for their permission to publish an English version of those papers and to receive the right to license all translations from this English version. This volume includes selections from the proceedings volume and a further development of a number of the essays, as well as additional material.

The Series Editors wish to acknowledge the support and encouragement of The Foundation of Thanatology in the preparation of this volume.

PREFACE

A United States–Japan Bioethics Congress was held in Tokyo, September 2nd–4th, 1994. A proceedings volume entitled *The Dignity of Death*, published in Japanese by Shibunkaku Publishing Co., in Kyoto, appeared April 26th, 1995. It is now a great pleasure to present to an English speaking audience the fruits of our Congress, together with some additional commentaries, as part of the Philosophy and Medicine book series.

My original purpose for organizing the conference was to explore transcultural comparative bioethics between the United States, Canada, Germany and Japan. Modern bioethics has its impetus and origin in the patients' rights movement of the 1960's in the individualistic democratic society of the U.S.A. The movement led to gradual changes in traditional physician-patient relationships. When Western bioethics was introduced into Japan, it met with some resistance and much hesitation. It was not easy for the Japanese to comprehend the need for Western principles of bioethics, such as informed consent, since the Japanese people have no urge to demand patient rights. Japanese society retains a traditional unease about democratic individualism and maintains collectivistic and paternalistic tendencies. Importance is still attached to unified social consensus. Journalists, for example, point to a lack of social consensus on issues concerning brain death. They claim that there is no consensus in Japan on whether brain death is socially accepted as the death of a person, even though it is understood to be a type of medical and biological death. As a result, no organ transplantations from brain dead donors have been undertaken in Japan for more than twenty years.

There are many subtle and overt racial, national, social, cultural and religious differences among divergent societies, such as Japan and the United States. Such differences may explain the difficulties that the Japanese and other cultures have in accepting many Western principles of bioethics. It may, in fact, be unethical to force people living in very different societies to accept Western bioethics. A truly international dialogue is necessary to overcome these types of problems. As a result, the U.S.–Japan Bioethics Congress was planned, organized and convened in 1994. Much of what was discussed and learned is presented in this volume in the hope that it will spark new and invigorating international bioethical debate.

Kazumasa Hoshino
International Bioethics Research Center
Institute of Religion and Culture
Kyoto Women's University
Kyoto, Japan

H. TRISTRAM ENGELHARDT, JR.

JAPANESE AND WESTERN BIOETHICS: STUDIES IN MORAL DIVERSITY

The essays in this volume, while exploring bioethical issues concerning death and dying, the use of scarce resources, and genetic interventions, also implicitly, and at times explicitly, compare Japanese and Western approaches to bioethics. As this volume shows, there are similarities. There are also marked differences. The similarities reflect a common attempt outside of any particular culture to find morally justified bases for collaboration when individuals do not share taken-for-granted understandings of the proper use of health care and the correct ways to collaborate. Both Japanese and Western scholars recognize the challenge of articulating a moral framework when they do not find themselves part of a particular moral community sharing a particular content-full morality. Similarities derive as well from content-full Western moral understandings that have been exported to Japan and which for better and worse have changed Japan. They also derive from Japan's exposure to the post-traditional character of our age. Many of the dissimilarities between Japan and the West stem from the circumstance that Japan still remains in great measure a traditional society with strong commitments to family and community. Japanese still share many common understandings of values. The dissimilarities stem in addition from the impact of the West's long struggles with its own moral diversity.

These introductory reflections concerning this volume's essays are developed against the significant background character of the West: the West is in the process of recognizing and addressing its fragmentation into a plurality of moral communities with a diversity of moral visions. The 16th century shattered the Western religious understanding into numerous Protestant religions on the one hand, and the Roman Catholic on the other. This rupture invited a philosophical attempt to bridge the resultant moral diversity through a general, rationally justifiable morality. If religion separated, reason would unite. The modern philosophical

Kazumasa Hoshino (ed.), Japanese and Western Bioethics, 1–10.

and Enlightenment reaction to the cultural and religious diversity that followed the Reformation, as well as to the bloody wars the Reformation fueled, was to search for a philosophical foundation for a secular equivalent to the dominant monotheistic morality the West had produced during the Middle Ages. This monotheistic morality, which defined the Western worldview, began to emerge in the 8th and 9th centuries and took substantial shape in the 13th. Against the bloodshed of the Thirty Years' War and the Civil War in Britain, the hope intensified of finding in or through reason a justification for the general lineaments of that morality so as to provide a non-religious moral foundation for political institutions, public policy, and social structures.

This hope of grounding the content of morality through rational argument appeared well-founded. At the end of the 8th and the beginning of the 9th centuries, as the West began to fashion its own Christian religion in contrast to the prevailing Christianity of the Mediterranean basin, it began as well to tie its religious faith and morality to philosophical reflection (Romanides, 1981). In the process, Western moral thought became allied with elements of the scholasticism that flourished in the High Middle Ages of the 12th and 13th centuries. In particular, Western moral thought embraced assumptions regarding the rational disclosability of the moral law. Despite the skeptical and nominalist reactions of the 14th century as well as the rejection of scholasticism by the leaders of the Reformation, this commitment to reason endured (Buckley, 1987). Indeed, as Western Christianity fragmented into numerous religions, sects, and cults, the philosophical project of a rationally justifiable ethics assumed greater importance.

Beauchamp's essay reflects this Western attempt philosophically to discover and justify a common, content-full morality. Though his claims for philosophy's foundational capacities are qualified, his exploration of Japanese and American bioethics presupposes the availability of a common morality that has content without being the property of a particular culture. He advances a common morality for a common bioethics that forms the basis for a moral community binding all mankind. In contrast, Hoshino offers an account of a particular bioethics embedded in a particular culture: Japan's. In accenting the ways in which Japanese patients regard themselves as members of social units, Hoshino shows the plausibility of interpreting issues of autonomy, confidentiality, and privacy within the context of the family. By underscoring that moral meaning is contextual, he speaks to a characteristic of morality that holds across cultures. By indicating the central place of Wa in Japanese reflections, with its endorsement of submission, he signals a difference between traditional Japanese cultural assumptions and those within which most Western bioethics is articulated. Here, Japanese law may better mirror Western assumptions than traditional Japanese values and moral commitments. Japanese law may have incorporated Western moral and societal perspectives that then fit only with difficulty onto traditional Japanese accounts of appropriate deportment. This difference in perspectives may portend an impending fragmentation of Japanese society into a cluster of traditional moral understandings as well as into a cluster of Westernized moral understandings, each bringing its own genre of bioethics.

The exploration of caring, compassion, and affect by Joy Penticuff offers an alternative Western moral approach to capturing content for bioethics. The challenge is again to gain content without losing universality. As Hoshino's remarks imply, if moral content is contextual, then caring, compassion, and affect can have no particular content outside of a particular socio-cultural context, outside of a particular moral community. This difficulty is cardinal. Hence, the quite different roles of rituals in Japan and the West, especially in the United States. Japan is a country where moral context and content are guarded by substantial common expectations and generally accepted religious rituals. Penticuff's pleas originate in a culture in which such common rituals are in decay or, when substantive, tend to be divisive. As a result, what it means to care and respond emotively to a situation will differ when individuals are separated by moral communities, religious commitments, and different understandings of what is ritually appropriate.

A sense of the distance between traditional Japanese moral understandings and those of the West is provided by Namihira's explorations of Japanese thoughts and feelings with regard to the remains of the deceased. These explorations offer a fine-grain map of the interplay of ritual, aesthetics, etiquette, and moral meanings. The result is a poignant contrast between the essays of Namihira and Penticuff. For the Japanese, there appear to be widely shared expectations concerning care for the dead. Japanese ritual expectations contrast with those in the West, which usually divide accessibility to rituals along confessional lines. Generally, it is a violation of religious moral expectations if Jews are buried according to Christian rituals, Christians according to Jewish rituals, Protestants according to Roman Catholic rituals, Roman

Catholics according to Protestant rituals. There are even significant divisions of expectations regarding funeral rites among the various Protestant religions. Such expectations often continue to govern, even if the families of the deceased are quite secularized and regard their particular religion's requirements more as cultural than religious obligations. The West's fragmentation into diverse religious-moral communities has produced barriers to rituals. Again, rituals divide rather than unite.

The papers by McKenny and Wildes provide further portrayals of the fragmentation and change in Western moral traditions. They also show some of the implications of these processes for bioethics. McKenny's essay addresses the post-traditional character of American society in which physicians have ceased to be the authority figures they were in the 1950's. The absence of a guiding traditional moral framework has led to an aesthetic of individuality along with an often disorienting anomie in the face of life-and-death decisions. Bioethicists have entered this authority vacuum, often functioning as if they could give guidance by an appeal to an unambiguous canonical morality. Yet their moral guidance is unable to set convincing secular limits or to provide general secular content-full direction in the face of mortality. Where Japanese religion still provides tradition and content, America's post-traditional society offers fewer resources to address limits in the face of death. Though Americans often aspire to a communitarian ethic, they remain in great measure irredeemably individualist in their search for moral community. Japan may have greater resources to embed concerns regarding death and dying in traditional Buddhist understanding than much of contemporary America has to recast its concerns in traditional Judeo-Christian religious moral commitments.

Wildes reenforces these observations about the context-determined character of moral commitments regarding death and dying by showing that terms such as "sanctity of life," which have played central roles in recent American bioethical debates, can only have force and unambiguous meaning within a particular moral community. The difficulty is that America, and the West generally, have fragmented into numerous moral communities: traditional Western Christian arguments are no longer able to guide general moral reflection and analysis in bioethics. Those who appeal to notions such as sanctity of life or human dignity in fact appeal to diverse notions interpreted within diverse contexts, inviting a chaos of misunderstandings rather than one coherent understanding (Bayertz, 1995). Moral babble follows the injudicious invocation of strategically ambiguous moral concepts. The result is a serious intellectual challenge. For example, as one approaches quality-of-life decisions, as Keyserlingk's explorations suggest, fundamental ambiguities disclose a rich fabric of different meanings of quality-of-life behind what might appear to be a unitary concept.

At the very least, the moral traditions framing bioethical expectations are quite different in Japan versus the United States. Moreover, there appears to be a greater likelihood of more agreement in more circumstances in Japan. This observation is strengthened by the exploration of the place of autonomy and communitarian approaches in terminal care offered by Veatch. Veatch's heuristic invocations of community also underscore the cardinal problem: the recurring difficulty of specifying which community and which communitarian values are to give guidance in bioethics and health care policy. All concrete expressions of community and autonomy are, as Hoshino underscores, contextual. The problem is to specify the context in which particular bioethical decisions are being made. Clarity in this regard requires specifying the moral community within which the moral context gains substance. This is an increasing challenge for Japanese as well.

Baruch Brody's careful analysis of futility further reenforces the contextual character of bioethical decisions. Brody assesses the moral warrants for limiting the desires of physicians, as well as patients and families, who might wish to fight against death to the end. As Brody shows, drawing on arguments from Francis Kamm (Kamm, 1993), the sense and importance of postponing death will be a function of how one understands death to be bad and life to be good. Brody invokes Kamm's reflections concerning the deprivation, insult, and extinction factors that make death bad, as well as the experience and achievement goods that make life good. As Brody argues, such analyses can help clarify how particular understandings of the badness of death and the goodness of life can make accepting death appropriate in some circumstances but not in others. Although Japanese understandings of proper medical decision-making and of appropriate responses in the face of death may show many similarities, especially when compared with their analogues in America, there will be a diversity of understandings in Japan. Hinohara's attention to the diverse roots of Japanese attitudes towards death, drawn as they are from Shintoist, Buddhist, Taoist, Confucianist, and Christian roots, is very helpful in disclosing the often unnoticed pluralism in Japanese culture. Moral diversity has implications for both Japanese and Western bioethics.

Appreciating what such moral diversity might mean for bioethical decision-making in Japan may be more difficult than it appears at first blush. As Yamazaki observes, physicians make decisions involving important values, even though it may be far from clear which values should guide such decisions. Even in Japan there will be the need, as Akabayashi's reflections suggest, to understand the cultural ambiguities that frame experiences of death and happiness, that shape the key terms through which medical decisions are framed. This may require in many areas substantially revising Japanese practices regarding consent, so as to recognize those persons who would want an American style disclosure of information. In addition, as Ohi stresses, not only are there differences between first-world and non-first-world expectations regarding privacy, technology, truth-telling, and the role of communities, but there are also differences within the first world and the developing world. In both, one encounters communities more or less embedded in their traditional moral and cultural understandings. Again, this is the case with respect to countries such as Japan that have maintained substantial elements of their traditional culture. For example, in Japan, one encounters tensions between traditional moral understandings and understandings of proper conduct that have taken shape through contact with the West. Similarly in the West, one finds tensions and conflicts between traditional and post-traditional moralities, between traditional moralities and emerging contemporary secular moralities.

The essays by Bayertz and Engelhardt explicitly and implicitly explore these differences and tensions within American and Western European bioethics as these address moral hesitations regarding non-coercive eugenics and the peaceable enhancement of human capacities. The issue of peaceable approaches to non-coercively improving human capacities contrasts starkly with the issues raised by the atrocities committed in this century against the innocent by secular regimes in first-world countries in the name of human improvement. The immorality of those undertakings can be appreciated in terms of the fundamental rights of individuals not to be used without their permission. It is easy and proper to condemn the violent and coercive eugenics of the Fascists who acted against individuals without their consent. It is quite a different matter to articulate in general secular terms hesitations regarding the improvement of human capacities and the enhancement of human abilities that can be undertaken without coercion, perhaps even with the enthusiastic agreement of all the moral agents directly involved.

How does one within a general secular framework articulate grounds to forbid a eugenics undertaken freely through germline genetic engineering? Concerns in principle regarding the consensual use of genetic technologies to enhance human abilities are difficult, perhaps impossible, to express in general secular terms (Engelhardt, 1996). Hoshino's reflections on the challenge of maintaining content-full Japanese notions of character, virtue, and community are again relevant here. Any account of the morality of the uncoerced enhancement of human abilities will need to be articulated within the context of a particular content-full bioethics. This is well illustrated by the European, and in particular German, attempt to assign a special normative status to the human genome, so as to proscribe a range of technological interventions in reproduction as well as the genetic enhancement of human abilities. Given the post-traditional character of Western European culture, these moral concerns may reflect a hunger after moral content and a nostalgia for the morality of a presecular past. Bayertz and Engelhardt in different ways address the problem of giving support for such proscriptions outside of a particular moral tradition. The difficulty is that, outside of a particular moral tradition, in the absence of framing accounts and generally agreed-upon sustaining justifications, proscriptive intuitions regarding such substantive choices can appear as merely unfounded hesitations. Like the 19th century Hawaiians whom Alasdair MacIntyre (MacIntyre, 1990, pp. 180–190) considers, moral intuitions deprived of supporting structures to give them sense and coherence become mere taboos. In such circumstances, moral debates are inconclusive, both within post-traditional societies whose moral practices are no longer intact, as well as across moral communities, which fail to share common moral premises, rules of evidence, and rules of inference.

The concluding two essays in the volume take stock of this collection as a whole. Ruiping Fan addresses the possibility of cross-cultural explorations of morality and bioethics, a sub-theme of this volume. By reference to the essays in this collection, he advances a number of methodological concerns regarding a comparative bioethics, in particular, justifying the application of one's own moral principles to a foreign culture and its bioethics. As he underscores, one must take cross-cultural moral differences seriously. In this regard he stresses the problem of determining: (1) whether there are differences in the moral vocabulary and fundamental moral principles employed in the bioethics of different cultures; (2) whether moral principles used in the bioethics of different cultures that appear to be similar are in fact divergent; and (3) whether there is the same ranking of principles and goods when the same ones are invoked in the bioethics of different cultures. The concluding essay by Mark Cherry takes stock of the post-traditional context that encumbers bioethics in America, Europe, and Japan. As the essay shows, although a greater moral and aesthetic integration may exist within Japan, it, too, has been touched by the challenge of post-modernity. The significance of the moral diversity that post-modernity brings depends very much, as he argues, on whether we are confronted with what he terms a metaphysical or merely epistemological relativism. The significance of our situation depends very much on whether moral diversity is grounded on a failure to know truly which moral principles and values should bind, or whether the diversity is due to the very absence of such principles and values.

As the essays in this volume show, there is a rich array of similarities and differences among the bioethics in America, Europe, and Japan. Many exploring bioethics and health care policy reveal a common hunger after moral content, as well as the community-specific character of such content. Many seek content-full, canonical, universal secular moral truths, while at the same time revealing the difficulty of rationally justifying claims regarding such truths. There are differences in the essays reflecting the extent to which the essays are framed within traditional moral assumptions or within those of a post-traditional moral context. The second differ in the degree to which they hold that moral content can still be secured in a view from nowhere outside of any particular moral tradition or recognize that such a project cannot succeed.

Japan and the West face similar challenges, but with different resources. On the one hand, the West has a longer and in some senses much deeper tradition of philosophical analysis and discursive argument, which has expressed itself in a well-articulated discipline of bioethics. This discipline, often carrying with it very particular moral and metaphysical presumptions, has at times been uncritically exported to the rest of the world, including Japan. Both the exporters and importers frequently fail to recognize the extent to which the character of that bioethics reflects the character of the Western predicament: the West is in great measure post-traditional, fragmented into often starkly different, indeed polarized, moral communities. On the other hand, Japan may still in many respects be more of one moral community rather than one society spanning numerous distinct and diverse moral communities. Yet, there already exist in Japan indications that it is entering a post-traditional era. Though it may in many respects appear to be one community, Japan fails sufficiently in this regard so that it is now confronted with many of the public policy challenges engendered by the moral pluralism confronting the West. In particular, Japan is now in need of secular moral resources so that Japanese moral strangers can communicate with each other and collaborate with general secular moral authority.

Against the background of these reflections, Japanese scholars and health care policymakers can recognize a brace of challenges. On the one hand, they must take account of the ways in which Japan, like the West, is becoming a post-traditional, morally fragmented society. This process of fragmentation will change the ways in which Japanese health professionals and health care institutions now and in the future should behave towards patients and the Japanese public generally. On the other hand, Japanese bioethicists have an opportunity to articulate a Japanese bioethics. Japan in great measure lacks a thoroughgoing attempt to fashion a Japanese bioethics: an understanding of its own traditional values and moral commitments with regard to bioethics. In this volume, Hoshino's reflections are very foresightful. They diagnose a cultural vacuum and suggest how to articulate an appropriate response.

A Japanese bioethics would be one that takes seriously traditional Japanese moral and aesthetic concerns and shows their relationship to the provision of health care and the framing of health care policy. The articulation of a Japanese bioethics could make a significant contribution both to Japanese culture and to Japanese public policy by helping Japanese to understand better the ways in which Japanese health care professionals and institutions tend to behave differently from those in the United States and the West generally. Insofar as a Japanese bioethics can be placed within still robust Japanese traditions and moral practices, many Japanese may be able to avoid having their moral commitments become like the taboos of the Hawaiians and their moral intuitions like the moral sensibilities of isolated American individuals who hunger after a community and moral content, which they have either abandoned or which no longer exist. Japanese will then have within their own society a counterpoint rich in their history from which to meet the moral diversity even they must take seriously and address.

If a Japanese bioethics or set of bioethics is not articulated, Japan will have lost the opportunity to have a special cultural resource through which critically to assess what it should accept from the West, to what extent, and under what circumstances. Japanese bioethicists can contribute by helping Japanese distinguish between that element of bioethics from the West that provides strategies for the peaceable collaboration of moral strangers, and those elements of bioethics from the West that reflect particular Western moral commitments, which at present are also often in some confusion and disarray. As Japan fashions its own bioethics, it will have much to contribute to an emerging international dialogue concerning bioethics and health care policy.

Center for Medical Ethics and Health Policy
Baylor College of Medicine/Rice University
Houston, Texas, USA

BIBLIOGRAPHY

Bayertz, K.: 1995, *Sanctity of Life and Human Dignity*, Kluwer, Dordrecht.

Buckley, M.J.: 1987, *At the Origins of Modern Atheism*, Yale University Press, New Haven, Conn.

Engelhardt, H.T., Jr: 1996, *The Foundations of Bioethics*, 2nd ed, Oxford University Press, New York.

Kamm, F.M.: 1993, *Morality, Mortality*, Vol. 1, Oxford University Press, New York.

MacIntyre, A.: 1990, *Three Rival Versions of Moral Enquiry*, University of Notre Dame Press, Notre Dame, Ind.

Romanides, J.S.: 1981, *Franks, Romans, Feudalism, and Doctrine*, Holy Cross Orthodox Press, Brookline, Mass.

PART ONE

ETHICAL UNIVERSALITY AND CULTURAL PARTICULARITY

KAZUMASA HOSHINO

BIOETHICS IN THE LIGHT OF JAPANESE SENTIMENTS

Western bioethics was introduced to Japan more than a decade ago. It seems obvious, however, that crucial problems exist in the acceptance of Western bioethics by the general public of Japan. Japanese people in general seem to be insensitive to or even subconsciously reluctant to accept the vital importance of autonomy, self-determination and individualistic freedom in decisionmaking; all of which are indispensably valuable principles in Western bioethics.

I. REFLECTION OF JAPANESE SENTIMENTS ON LEGAL STATUS IN JAPAN

In 1979, the "Cornea Transplant Act of 1958," which was the sole legislation on organ transplantation in Japan at the time, was revised. The "Cornea and Kidney Transplant Act" was enacted. Surprisingly, however, this new act was confined to cornea and kidney transplants from dead donors. Organs may only be removed from the donors with the written consent of the surviving family. This is the case even when the deceased individual expressed in advance the desire to donate. Moreover, no specific hierarchy in the surviving family members is established for providing consent to the organ donation. As I have been well familiarized with the Uniform Anatomical Gift Act since it was enacted in 1968 in the United States, it was very difficult for me to accept this new act without emotional uneasiness. Western bioethicists might believe without any hesitation that the wish to donate a cornea or kidney expressed in advance by the deceased must be one of the advance directives in Japan. This written wish, however, is not a legalized advanced directive. In Japan, thus far, no advance directives nor living wills for health care have ever been legalized. It seems most probable that Japanese people are in general not inclined to have the active motivation for legal protection of their own rights to autonomy,

Kazumasa Hoshino (ed.), Japanese and Western Bioethics, 13–23.

self-determination, living-wills and other advance directives. In fact, the concept of an advance directive, including a durable power of attorney for health care decisionmaking, is not well understood by the general public in Japan.

One year after the enactment of the "Cornea and Kidney Transplant Act of 1979," I was appointed the Chairman of the Committee and assigned to lead an energetic nationwide campaign to legislate a new act for the donation of the whole body for anatomical studies at the medical centers in Japan. Our efforts brought us success in legislation. New legislation, "Act of Donation of the Body for Medical and Dental Education" was at last enacted on May 25, 1983. Regrettably, however, we failed to convince the Diet members of Japan of the importance of the clause indicating that the advance directive by the deceased must be fulfilled. Thus, identical to the "Cornea and Kidney Transplant Act of 1979," the advance wish of the deceased must again be agreed to by the written consent of the surviving family.

II. AWARENESS OF PATIENTS' RIGHTS

While I was working in Japan, I began to realize that most Japanese patients and their families do not seem to be aware that patients have rights. Most Japanese patients tend to think that they are asking a favor of the physician to take care of their medical problems. They, thus, try to be as good and obedient a patient as possible. Generally, patients in Japan do not care much about their civil rights, except for those individuals who are specialists or members of a social movement. Contrary to Americans, who undertook the "Patients' Rights Movement" in and following the 1960s, the Japanese have never had inevitable urges for a similar movement. They have passively received the principles of bioethics from the United States.

Sometimes patients in Japan become annoyed when physicians attempt to explain medical conditions and treatment procedures in detail, because they think that these professional matters should be taken care of by the physicians themselves. The notion is that patients should not be burdened by such information and decisions, rather, they should just receive good treatment and care. Patients are concerned merely about whether or not treatments that they are about to receive are painful and whether the results and consequences will be satisfactory to them. Once they decide to ask their physicians to take care of them, they tend to leave everything to the physician. There is a famous Japanese proverb, "manaita no ue no koi," which means that the patients are resigned to their fate. This type of decisionmaking by a patient is rather common and may be interpreted as a type of autonomous decisionmaking. The patient is autonomously resigning themselves to the physician's care.

Analogously, at any non-Westernized restaurant in Japan, once Japanese guests order a dinner course, they do not care about what other kinds of dishes are served to them. Customarily a waiter serves a salad with a certain kind of dressing without asking the choice of the guests. Japanese guests do not usually pay any attention to this paternalistic manner. But in my case, even now, it causes emotional unease. My friends remind me that I must become at ease with the Japanese custom. I had, then, to realize that the Japanese interpretation of this waiter's manner is quite different from my own impression. The restaurant's chef decides on behalf of the guests which dressing should be served with the meal in order to avoid interfering with the cozy atmosphere at the dinner table. In other words, Japanese guests accept this type of treatment as a part of the courtesy of the restaurant.

III. TRUTH TELLING TO JAPANESE PATIENTS

It has been well known that in the United States more than 90% of patients have the truth disclosed to them by their physicians regarding their medical prognosis, unless the patient specifically refuses to receive the truth. In Japan, however, this occurs with less than half of the patients. Different from individualistic Western societies, the general public in Japan usually does not appreciate the value of the right to know the truth of one's own psycho-physical medical condition. Needless to say, there are also individuals who are exceptions to this rule.

Some individualistic persons, particularly in the younger generations, criticize this tendency not to reveal the truth. As the patients mature and grow older, or as terminal stages of their illness are seemingly approaching, however, most Japanese patients tend to become nervous about receiving the final announcement of frightening and hopeless news about their severely sick conditions. Contrarily, many healthy people seem to think that truth-telling to patients is acceptable particularly when they consider the quality of life of the patients during their terminal stages. Once these individuals become seriously ill, though, they become unsure as to whether they really wish to have the truth divulged or not. Often seriously ill patients feel that it would not be of any use even if they learned the truth in detail, because complete recovery is no longer a real option. Why would it be worthwhile to hear depressing bad news while in their terminal state?

It is thought that it would be better for patients in their terminal stages to spend the remainder of their time with even some vague hope that some miracle might happen to save them. They would rather be ignorant of their true condition rather than suffer from the emotional upset of hopelessness. This type of psychological reaction is rather common among severely sick patients in Japan.

In general, Japanese people are not accustomed to making medical decisions regarding their own diseases by themselves without consulting the family. This is because of their deep regard and respect for the opinions and feelings of the family. When one member of the family becomes sick, it is the responsibility of the entire family to look after him. The main person among family members who looks after the sick individual is usually a female member who takes care of the household on a daily basis. She is inevitably over worked by the unexpected additional duties of caring for him, although other members of the family will typically try to share some of the work load and other household responsibilities. Thus, while she will make most of the sacrifices, others will help her as much as possible. The family knows that the care of the sick member is a family matter.

In these circumstances, it seems rather natural for the family first to decide on the best medical procedures and to care for him. The family usually asks what medical procedures and care he prefers to receive and at that time he has a chance to express his own ideas and wishes to them. In the situation where there is no disclosure of the truth about the sickness to the patient, it is difficult for both the family and the patient to have fruitful discussions. On such occasions, the patients typically leave the final decision to the family so that the family may take into consideration not only realistic medical factors but also the family's own convenience.

If it is decided to care for the patient at home, the normal daily life of the entire family is significantly disturbed and it creates a considerable burden on them, particularly when the care is for a prolonged period. Even when a patient is hospitalized it is not easy, because in most hospitals in Japan a considerable amount of care must be given to the patient by his family. A family member typically stays overnight in the ward or commutes daily to the hospital. The family usually must deal with financial matters as well. Unless the sick individual is paternalistic at home, he may soon realize that he is a burden to the family. Even when he is well off financially, he may nonetheless need to depend upon the family. Eventually, decisionmaking for medical procedures and care for the patient may be done with the mutual consent of both himself and the remaining members of the family.

As the responsibilities taken by the family become increasingly heavy, the family begins to consider it their duty to understand in detail the whole truth of the patient's prognosis. This duty is perceived without any thought that they might be infringing upon the patient's right to autonomy, self-determination and privacy. There is, in fact, no concept in the mind of the Japanese that one of the family members has such individual rights independent of the family. Japanese people, in general, tend to feel emotionally that the family must concern all private matters of one's own family members as if it were a family secret. They try to conceal private family matters from outsiders. Therefore, the family does not think of it as an infringement of privacy even when they have obtained private medical information directly from the physician without the patient's consent.

IV. JAPANESE IDEAS OF THE FAMILY UNIT

There has been a long tradition in Japanese society of "group" consciousness ("mi-uchi ishiki") which is accompanied by a strong sense of belonging to a group with affection and loyalty. Groups come in various sizes. Anything from a unit of one's own family, a clan of relatives, a group of close friends, a group of intimate colleagues, or even an entire company can be considered a cohesive group. There is a strong tendency for group members to distinguish themselves as the insiders from the outsiders or strangers. They often disregard the outsiders and treat them coldly, and sometimes even behave cruelly towards them.

When someone attempts to become a member of the "mi-uchi," this person must be accepted by the group, for example, by marriage to a member of the family. When a new member, say a bride, is added to a family consisting of five members, the average American may think that she joins as an independent person with her own identity and that she would be allowed to act on her own autonomous decisionmaking. The family would then consist of six separate individuals. Japanese people, in contrast, most probably think that a bride is blended in with the other family members. The bride usually tries her best to learn the lifestyle and running of her new household, and tries to intermingle with other family members harmoniously without revealing her own identity distinguishably. She must not behave like a guest of the house or an intruder. As a result, even after the bride becomes a new family member, the characteristics of that family unit remain basically unchanged. In the Japanese traditional wedding a bride wears a pure white wedding kimono suggesting that she will be dyed with any color that is the color of the family unit to which she joins as a newcomer.

A family member, therefore, thinks that all family members are a kind of extension of each other within the same unit. This type of logically unexplainable national sentiment may be one of the many reasons why the Japanese family often cannot agree to make their deceased member an organ transplant donor.

Each member of the family seems to be connected individually with each of the other members perhaps not only by love or affection as seen in individualistic society, but also by a certain kind of spiritual tie in this collectivistic society of Japan. This type of spiritual tie appears to be based upon the long lasting traditions, culture, and ethos of Japan. The nature of Japanese sentiment created by the spiritual tie of a daughter with her mother may well be different from that with her father or her grandmother. Also, the spiritual tie of an eldest son with his father, may change after the death of his father. Even when the son could not get along with his father, he must assume responsibilities as the head of the household. At times this experience allows the son to understand why his father behaved in the way that irritated him while his father was alive. Then it becomes difficult for the surviving son to sever his spiritual tie with his deceased father.

Particularly when the death of a family member occurs in unusual circumstances, such as combat, a plane accident, or brain death, the survivors find it difficult to accept the death of the deceased because their ties with the deceased remain strong. But then, their spiritual tie could be changed to make them feel much easier. This process may be one of the reasons why the surviving family in Japan hardly ever gives consent for removal of organs from the deceased for the purpose of organ transplant.

V. WA – A CONCEPT OF THE JAPANESE NATIONAL SENTIMENT

In order to understand the Japanese national sentiment, based upon long lasting tradition, customs and culture in Japan, it is essential to recognize the vital importance of the real meaning of the Japanese word "WA." The word cannot be translated adequately into an equivalent English word, although there are several English words each of which captures a small part of the comprehensive meaning. These include: conciliation, concord, unity, harmony, submission, and reconciliation. An intimate integration of all of these words may begin to resemble the real meaning of "WA."

Japanese social systems are based on either familialism, collectivism or totalitarianism, depending of the size of the group. The Japanese are usually uneasy or uncomfortable when they try to begin something if no unified opinion is formed in the group. In Japanese society, people are usually not encouraged to make autonomous decisions because everybody in a group is expected to follow the group opinion. Usually, the group opinion is not determined by any democratic means but by the dominate opinion expressed by the leader of the group, regardless of age, or by a person in a superior position. The insistence of one's self-determined opinion expressed particularly by inexperienced junior persons is disliked and discouraged by the groups. In this sense, Japanese society remains basically feudalistic.

Very often a certain conclusion regarding the agenda of the meeting is predetermined by the selected members of the group by "behind the curtain negotiations" before the formal meeting. Thus, in fact, the formal meeting is merely a sort of ceremony in order openly to make a final decision in a formal setting. This is a rather common Japanese custom in negotiation called "nemawashi."

An ordinary Japanese person would tend to think that it would be unwise for him to speak up with a personal opinion in a serious discussion. He would try to follow obediently the opinion decided upon at the meeting. The Japanese proverb says "nagai mono niwa makarero" which means that if you cannot beat them, you should follow them obediently. Nevertheless, individuals who at one time obediently accepted the group consensus at the formal meeting often sneakingly complain about it later. In contrast to the American expression "a squeaky wheel gets the grease," in Japan the maxim is "the nail that sticks out gets pounded down."

It is this group decisionmaking ethos which is why most Japanese people are not used to quick autonomous decisionmaking not only in regard to their own medical care but also in their daily lives. Consequently, the Japanese tend not to be sensitive enough to the other's rights to autonomous decisionmaking and they override other's individual rights without realizing that they are actually behaving unethical in the American sense. For example, when someone disagrees with an organ transplantation from a brain dead donor, this person is putting more weight on the lack of social consensus, than on the life of a fatally ill cardiac patient whose life may be saved by cardiac transplant.

This type of Japanese national sentiment may be criticized as being inhumane from the Western humanistic standard. There are, however, distinct differences in humanism between Western societies and the Japanese society. Native Japanese people are basically loyal as well as humane to the people in one's own "mi-uchi." In contrast, they are rather cool or neglectful toward outsiders or strangers. Group consciousness exemplifying WA dominates personalized individualism.

VI. THE INFLUENCE OF WA ON ART AND CONVERSATION

The beauty of Japanese flower arrangements may be created by unity and balance among asymmetrical and unbalanced objects. No insistence of self-assertion and self-reliance is expressed by any object. Yet, every object in the flower arrangement lives in unity with nature. This is also true in Japanese landscape gardening. There is complete harmony with self-renunciation and self-denial. Each object when it is looked at alone, however, expresses individually its own beauty and dignity.

In daily Japanese conversation almost no individual pronouns, such as I, my, or me, are used. This serves to minimize the insistence on self and individuality. Most Japanese are accustomed to speaking in a round about fashion rather than in a direct straight forward manner. The Japanese consciousness has mastered the art of reading between the lines as well as sensing and conjecturing the other's innermost thoughts and feelings. Thus when spoken to in a round about fashion, Japanese people can fill in the blanks in the conversation and comprehend one another accurately.

Given this tendency, the communication of distressful news requires even more courteousness and compassion. The Japanese physician would hesitate to disclose any distressful news directly to the face of a patient, especially without having already alluded to it in some roundabout fashion. Most probably, the physician would discuss the prognosis with the family first, but also with great care and indirectness.

In this indirect fashion of communication, conversations gently taper off rather than ending with a definitive statement. Especially with a touchy subject, care is taken to ease into an unrelated topic so as to reduce the strain of ending the conversation. One very important aspect of Japanese conversation is that neither party should mention anything which gives a person a "coup de grace," a finishing blow, which definitively closes the conversation up to the definitive final words completing an unpleasant story.

For instance, after a lengthy detailed explanation of the symptoms of terminal colon cancer, the patient may be able to sense how critically ill he is and does not want to have the doctor provide a "coup de grace." While many Westerners would appreciate the physician for his thorough consideration, most Japanese do not like to be told such definitive final news. Speaking with careful precautions in these types of circumstances is considered to be both polite and sophisticated in Japan.

VII. CULTURAL INFLUENCES ON BIOETHICAL APPLICATIONS

These characteristic ways of feeling, thinking and behaving reflect the national sentiment of the Japanese. They are strongly influenced by a number of environmental factors derived from the traditional Japanese culture and the long lasting traditional ethos. It would be extremely difficult to change the ethos of a country with theoretical arguments or by rational reasoning. Some of the fundamentally important principles of informed consent, such as autonomy and self-determination, are not readily accepted without conflict within the social environments of a country like Japan, England (Schwartz and Grubb, 1985), or southern Italy (Surbone, 1992).

Pellegrino has written that culture and ethics are inextricably bound to each other (1992). Culture provides the moral presuppositions and ethics for the formal normative framework of our moral choices. Every ethical system, therefore, is ultimately a synthesis of intuitive and rational assertions, the proportions of each varying from culture to culture.

Elliot has written that the practical difficulty with applying ethical theories to particular problems is that ordinarily people pay little attention to theories while they are making moral decisions (1992). Instead, we are guided by our ethical beliefs, which are primarily the result of cultural factors beyond our reach – factors subject to rational scrutiny and to changes, but largely outside of conscious control. Davis' expressed concerns are that collectivist cultures comprise about 70% of the world's population (1992). But virtually all of the data of modern psychology, and most other social sciences, come from the most individualistic of cultures, such as the United States. As a result, according to some social scientists, many Western assumptions about the universals of human behavior actually only apply to a minority of people.

VIII. CONCLUSIONS

A certain modification in the application of the principles of informed consent and others of Western bioethics to Japanese society without altering the fundamental values may well be needed. In the principles of informed consent and truth telling, for example, instead of emphasizing autonomous self-determination by the patient alone, the physician could perhaps ask the patient to have some member(s) of the family assist with the decision. Only through this type of alteration will Japanese patients and families begin to accept the practice of informed consent or other types of bioethical principles.

International Bioethics Research Center
Institute of Religion and Culture
Kyoto Women's University
Kyoto, Japan

BIBLIOGRAPHY

Davis, A.J.: 1992, 'Ethics of disclosure of information to dying patients – Problems with Western bioethics in the U.S.A.', in K. Hoshino (ed.), *Bioethics with Regard to Death of Patients*, Sokyusya Publishing Co., Tokyo, pp. 1–21.

Elliot, C.: 1992, 'Where ethics comes from and what to do about it', *Hastings Center Report* **22**(4), July-August, 28–35.
Pellegrino, E.D.: 1992, 'Intersections of Western biomedical ethics and world culture: Problematic and possibility', *Cambridge Quarterly of Healthcare Ethics* **3**, 191–196.
Schwartz, R. and Grubb, A.: 1985, 'Why Britain can't afford informed consent', *Hastings Center Report* **15**(4), August, 19–25.
Surbone, A.: 1992, 'Truth telling to the patient', *JAMA* **268**, 1661–1662.

TOM L. BEAUCHAMP

COMPARATIVE STUDIES: JAPAN AND AMERICA

Many today believe that secular pluralism has created a so-called post-modern world in which we should give up our robust past beliefs in the universality of moral precepts (Engelhardt and Wildes, 1994, esp. p. 136; Wildes, 1993; Engelhardt, 1986, pp. 39–103, 366–79). I will maintain, however, that a body of general ethical precepts constitutes morality wherever it is found. I will call this shared, universal system of beliefs "morality in the *narrow* sense." From this perspective, no difference exists in basic moral precepts between Japanese and Western morality. When the word "morality" is used exclusively in the narrow sense, it makes no sense to speak of fundamentally different *moralities* – as if Buddhist morality, Shinto morality, Jewish morality, and Christian morality were fundamentally different. However, different moralities are present in what I will call "morality in the *broad* sense," as I will discuss shortly. Morality in the broad sense recognizes divergent and even conflicting moral positions created by different philosophical, religious, or cultural commitments.

I will first defend this point of view by discussing the respects in which it can and cannot be said that contemporary Western and Japanese bioethics are relative to unique cultural norms. I will devote the bulk of this analysis to comparing and evaluating American and Japanese bioethics on themes of autonomy, paternalism, and informed consent. But first I will outline a few assumptions I will make about the nature of morality.

I. MORALITY IN THE NARROW AND BROAD SENSES

Not everyone in every society accepts or lives up to the demands of morality in the narrow sense. This is not because they have a *different* morality. Such persons are amoral or immoral. Morality, so understood,

Kazumasa Hoshino (ed.), Japanese and Western Bioethics, 25–47.

is not a philosophical theory or an ethical theory of any type. It is a group of unphilosophical precepts most often categorized as principles, rules, and rights. In recent years the favored category in international discourse has been universal human rights (Macklin, 1992, p. 245; National Committee, 1984; Gostin, 1987).[1] Since ancient times, many attempts have been made in European philosophy and law to show that moral precepts do not depend, as do mere customs and positive law, on local codifications. As Dutch jurisprudential thinker Hugo Grotius put it, these norms are obligatory for everyone and binding even on kings. They provide an objective basis for moral judgment and international law. There are many problems with these philosophical formulations, most of which were expressed through controversial ethical theory. I cannot here deal with these efforts, but I can provide some examples of rules of morality in the narrow sense, and I can say why this morality is universal. Here are examples of universal rules that all morally serious persons accept.

1. Tell the truth.
2. Respect the privacy of others.
3. Protect confidential information.
4. Obtain consent before invading another person's body.
5. Be loyal to loyal friends.
6. Do not kill.
7. Do not cause pain.
8. Do not incapacitate.
9. Do not cause offense.
10. Do not deprive of goods.
11. Protect and defend the rights of others.
12. Prevent harm from occurring to others.
13. Remove conditions that will cause harm to others.
14. Help persons with disabilities.

No rule on this list is absolute, and no rule is arranged or lexically ordered to override other rules in cases of conflict. Such ranking, if it were possible, is the work of ethical theory, and I am discussing morality, not ethical theory. This list is also not exhaustive of morality in the narrow sense. Other precepts such as fiduciary duties, obligations of reparation for wrongful interventions, obligations of gratitude for the generous services or gifts of others could be added. This list illustrates but does not exhaust morality in the narrow sense.

Far more social consensus exists about the acceptability of all these action-guiding precepts than exists about any philosophical, religious, or political theory of ethics. This is not surprising, given the central social importance of morality and the fact that its precepts are embraced in some form by virtually all major ethical theories. Ethical theories are rivals over matters of justification and method, not over principles of morality. The above precepts are at the deepest level of any person's commitment to a moral way of life, but any person with that commitment need know nothing about ethical theory to have that commitment.

The Need for Specification and Reform

David Hume once noted that, "the principles upon which men reason in morals are always the same; though the conclusions which they draw are often very different" (Hume, 1772). This pithy statement expresses an important truth: despite the nonrelativity of norms of morality in the narrow sense, a relativity or pluralism *of judgments and practices* is an inevitable outcome of moral decisionmaking and historical development in cultures. These judgments and practices constitute what I will call morality in the *broad* sense. Here the many conflicting social codes across societies show significant dissimilarities of belief, but not so many dissimilarities that morality in the narrow sense is called into question.

Why do we need to distinguish narrow from broad morality? One problem, and perhaps the main one, is this: the basic precepts of morality in the narrow sense are vague, general, and indeterminate precepts that can be implemented in many different ways consistent with their abstract content. Valid *exceptions* to the rules are also recognized in different ways in different cultures, groups, and individual decisions – thus introducing the phenomenon of *moral differences.*

The precepts of morality in the narrow sense usually provide little more than a basic orientation for addressing specific moral problems such as whether physician-assisted suicide is a morally licit enterprise. These practical moral problems require that we specify basic precepts and create policies and practices (Richardson, 1990). Bare, unimplemented, universal precepts, then, should not be expected to resolve any deep or complex moral problem, because we also need reasons for implementing the principle in one way rather than another. Every culture has gone through processes of making its morality suitably specific.

Rather obviously, not all cultures have given the same answers to the same problems, and often many answers have emerged from within the framework of a single larger culture such as that of Japan and the United States. In all cultures the interpretation and specification of norms, the reconstruction of traditional beliefs, the balancing of different values, and negotiation are essential on an ongoing basis. This perspective invites evolutionary change in ethics while insisting that morality in the narrow sense provides the starting point and the constraining framework. This is the heart of what happens in bioethics both as a philosophical enterprise and as a means to the formulation of institutional and public policy.

Because we *create* rather than simply *discover* our specified precepts, morality in the broad sense is "invented."[2] We invent the rules and policies that *specify* the commitments of morality in the narrow sense. For example, since approximately 1966 we have been inventing rules and policies in the United States to protect human subjects of biomedical and behavioral research. These rules extend beyond the content of any pre-existing principles and practices and beyond the substance of morality in the narrow sense. None of this should be surprising or even controversial. Both groups and persons typically lack a complete understanding of the full range of commitments they make in accepting a moral precept, because of its indeterminateness and our inability to specify the precept all the way down to the concrete cases. I am proposing, then, that in managing complex or problematic cases involving contingent conflicts, the first line of attack should be to specify pre-existing norms and thereby to reduce or eradicate the conflicts.

II. INFORMED CONSENT IN JAPANESE AND AMERICAN BIOETHICS

I now come to my main subject, which is a comparison of the American and Japanese contexts on themes of autonomy and medical paternalism.[3] I will confine attention largely to informed consent.

I take the following view of informed consent and its relation to morality: the obligation to respect the choices of autonomous persons runs as deep in morality as any of morality's precepts. Obtaining a person's informed consent is one way of respecting that person's autonomy. But no principle of respect or autonomy right is absolute, and for centuries of medical practice it was found convenient to override a person's autonomy interests in order to promote the person's welfare. The status of the goal of promoting the patient's best medical interests explains why informed consent was never, until recently, central to medical ethics, but this piece of history in no way severs the tight bond between respecting persons' rights of autonomy and respecting their rights to consent. This obligation to obtain a patient's or subject's informed consent is a universal and not merely a regional obligation, but it also is not necessarily an overriding obligation in all cases.

With these assumptions in hand, I will now argue that morality in the narrow sense is already recognized in both American and Japanese cultures and that their two recent histories of addressing problems of medical disclosure and consent are not as different as they have been depicted in the literature on this topic. Any differences, I will argue, are differences in degree rather than differences in kind. However, any differences in degree are present as much within these cultures as across these cultures.[4]

Informed Consent and Paternalism: American and Japanese

It is often said that informed consent is a peculiarly American phenomenon – or at least a notion that fits more naturally into a traditional Western ethics of individualism than an Eastern ethics of relationship in communities. From what I have said already, you can see that I am skeptical of this thesis. Although the practice of obtaining informed consent arose in the United States, the view that it is a peculiarly American or European phenomenon ill-suited for other cultures, including Japan, is an untenable thesis. I will begin to address this thesis by outlining what I will call *the received view* of the relevant cultural differences between Japan and the United States. I call it the "received view" because it pervades dozens of published articles comparing American and Japanese cultures and has no significant competitor in literature on the subject.

The Received View[5]

The received view is that physicians in Japan are typically paternalistic and authoritarian in their treatment of patients and families. Physicians believe they should make the critical decisions, while shielding patients and often families from the frustration, anxiety, and perplexity involved in medical decisionmaking. Patients and their families are correlatively deferential to physicians, willingly yielding to their advice and decisionmaking authority. Traditional family and religious values in Japanese society feed this custom: individuals are expected to be relatively constrained and unassertive, contributing to the maintenance of fluent relationships that avoid confrontation and self-assertion. This behavior and set of cultural expectations contrast sharply with the Western emphasis on individuality, which, according to the received view, is not admired in Japanese culture. For this reason, informed consent has not flourished.

Exponents of the received view maintain that the situation in Japan stands in marked contrast to views in the U.S. and that these differences spring from moral differences in the two cultures. As Rihito Kimura puts it, "Autonomy, an important bioethical principle in the Western social context, is out of keeping with the Japanese cultural tradition;" (Kimura, 1986, p. 23) and, as Norio Higuchi maintains, "Americans criticize Japanese for their paternalism, while Japanese counterattack Americans for too much autonomy without professional responsibility" (Higuchi, 1992, p. 457).[6]

I will now criticize this received view. I do not deny that this view is valid as an expression of differences *in degree* between American and Japanese society, but I believe it has no merit as an account of differences *in kind*. For example, it would be inaccurate to say that the Japanese medical environment is paternalistic, whereas the American medical environment is not. I need not, then, deny all credibility to themes in the received view, but I will argue that the received view is a misleading account of U.S. and Japanese dissimilarities and a thoroughly unilluminating account of what the role of informed consent should be in both cultures.

The American Situation

Since the early 1970s, informed consent has become a widely accepted doctrine in American law, bioethics, and biomedical institutions (Faden and Beauchamp, 1986). However, informed consent's history is short. The term "informed consent" first appeared in the U.S. during the 1950s. Prior to this time, there was little in the way of firm cultural grounding in which a commitment to informed consent could take root in the United States.[7] Moreover, serious discussion of the concept and commitment to its presence in medical and research practice began only around 1970. It therefore has no prolonged cultural history. In fact, the histories of consent and disclosure in the medical establishments of Europe and the United States are as antithetical to informed consent as they are hospitable. That history is primarily a history of physician control of information and cultural deference to physicians, much as in Japan. Still today patients, families, and courts in the U.S. remain highly deferential to physicians. Little evidence exists that, until very recently, requirements of informed consent had a significant hold on the practice of either physicians or research investigators.

Physicians' views in the U.S. about proper consent practices differed markedly as recently as the late 1960s from the conventions supposedly in place today. For example, in one study reported in 1970, half of the physicians surveyed thought it medically proper, and thirty percent ethically proper, for a physician to perform a mastectomy with no authorization whatsoever from the patient other than her signature on the blanket consent form required for hospital admission; similarly, more than half the physicians thought that it was ethically appropriate for a physician not to tell a cancer patient that she had been enrolled in a double blind clinical trial of an experimental anti-cancer drug (Hagman, 1970; see also Hershey and Bushkoff, 1969; Fellner and Marshall, 1970; Alfidi, 1971).[8]

If we reach back farther in history, the major writings of prominent figures in ancient, medieval, and modern medicine contain a storehouse of information about commitments to disclosure and discussion in medical practice. But it is a disappointing history from the perspective of informed consent. Throughout its history, European medical ethics developed predominantly within the profession of medicine. With few exceptions, no serious consideration was given to issues of either consent or self-determination by patients and research subjects. Proper principles, practices, and virtues of "truthfulness" in disclosure were occasionally discussed, but the perspective was largely one of maximizing medical benefits through the careful management of medical information. The central concern was how to make disclosures without harming patients by revealing their condition too abruptly and starkly. Withholding information and even outright deception were regularly justified as morally appropriate means of avoiding such harm. The emphasis on the principle "first, do no harm" even promoted the idea that a health care professional is obligated *not* to make disclosures because to do so would be to risk a harmful outcome.

I therefore find it difficult to accept the received view's thesis that Japanese traditions are deferential to physicians and rest on a paternalistic model of medicine, whereas American (and perhaps European) traditions insist on the rights of patients and rest on an autonomy model of the patient-physician relationship. For this depiction of American society to be credible, it must be confined to the last twenty-five years of American history.

Moreover, this depiction has little credibility as a description of the culture of American medicine even today. Although procedures of informed consent have taken a firm hold in many parts of medical practice in the U.S. – for example, routine practice encourages the obtaining of signatures on consent forms and the disclosing of information about alternative treatments, risks, and benefits – this hold is still very tenuous in many parts of contemporary American medicine.

Among the best data on this subject are the findings of a national survey conducted by Louis Harris and Associates in 1982 for the President's Commission for the Study of Ethical Problems in Medicine and Biomedical and Behavioral Research (President's Commission, 1982, Vol. 2). Almost all of the physicians surveyed indicated that they obtained written consent from their patients before in-patient surgery or general anesthesia. At least 85 percent said they usually obtain some kind of consent – written or oral – for minor office surgery, setting of fractures, local anesthesia, invasive diagnostic procedures, and radiation therapy. However, other evidence in this survey questions the depth and meaningfulness of this consent-related activity. In one question, this survey asked physicians, "What does the term 'informed consent' mean to you?" In their answers, only 26 percent of physicians indicated that informed consent has something to do with a patient's giving permission, consenting, or agreeing to treatment. In a related question, only 9 percent indicated that informed consent involves the patient's making a choice or stating a preference about his or her treatment (President's Commission, Vol. 1, p. 18; Vol. 2, p. 302).[9]

The overwhelming impression from this and other parts of the recent empirical literature on American practices and from reported clinical experience is that the actual process of soliciting informed consent often falls far short of the ideal of a serious show of respect for the decisional authority of patients (Meisel and Roth, 1983).[10] As the authors of a 1991 empirical study of American physician-patient interactions conclude, "despite the doctrine of informed consent, it is the physician, and not the patient, who, in effect, makes the treatment decision" (Siminoff and Fetting, 1991, esp. p. 817). Thus, although legal requirements of informed consent are reasonably settled in most American legal jurisdictions, the implementation of those laws is frequently faulty, and the underlying practices laced with paternalism.

The Japanese Situation

Is the situation significantly different in Japan – either as a matter of cultural attitude or medical practice? There are reasons to doubt that significant differences distinguish Japan and the U.S., which is *not* to say that there are *no* differences. To assess the situation in Japan, a constructive proposal made by Koichi Bai, Yasuko Shirai, and Michiko Ishii will be worth bearing in mind as we proceed:

> We must beware of drawing general conclusions as to "Japanese" characteristics on [bioethical] issues. We ought not to assume too readily a uniformity in Japanese culture; nor, needless to say, can we ignore Japanese peculiarity. The key is to observe the situation as it exists.... Japanese attitudes have not been examined closely because of misinformation and lack of research (Bai, Shirai and Ishii, 1987).[11]

To examine the available research, consider now three recent surveys of the opinions of Japanese physicians. The first was reported by Professor Hiroyuki Hattori of the Yamaguchi University School of Medicine and five of his associates and published in 1991 (but conducted in 1986–87), the second reported by Professor Yutaka Mizushima and eight of his associates and published in 1990 (but conducted in 1989), and the third reported by Ichiro Kai of the School of International Health, University of Tokyo, and six of his associates, and published in 1993 (but conducted between 1985 and 1988). I begin with the Hattori Study, as I will call it.

The Hattori Study suggests some striking similarities to the empirical studies in the U.S. about American physicians. This questionnaire survey concerning informed consent was administered to Japanese physicians in Yamaguchi prefecture. These investigators reached the following conclusions about the results of their study. "The survey showed that even though these Japanese physicians are willing to give their patients sufficient information to obtain informed consent, the discretion of the physician to provide information is still prevalent." The survey "revealed that Japanese physicians believe that information regarding the treatment to be administered should be fully disclosed both in cases when the treatment is still experimental and when it is established among specialists" (Hattori, Salzberg, Kiang, Fujimiya, Tejima and Furuno, 1991). In every category tested, over 50% of Japanese physicians stated their conviction that they should make disclosures consistent with informed consent. One of the most interesting responses came in answer to the question, "How do you explain high-risk diagnostic procedures to the patient?" The poll showed a consistent result across medical students and physicians in university and other hospitals of from 56% to 60% who responded, "we explain the incidence and the severity of the risk, and if the patient seems to be bewildered by the information, we explain them to the relatives" (Hattori, p. 1013). Again, these results are consistent with studies of the behavior of American physicians.

The Mizushima Study, as I will call it, examined disclosures of a diagnosis of cancer in Toyama Prefecture, and asked the opinion of physicians, para-medical personnel, and lay persons about Japanese practices of nondisclosure (Mizushima *et al.*, 1990, esp. p. 146). A major motivation for the study was the widespread belief mentioned in the opening paragraph of the study that, "[i]n Japan, more than 90% of medical doctors hide the actual diagnosis of cancer from patients. On the contrary, in the Unites States of America, . . . more than 90% of MDs reveal the diagnosis of cancer to their patients." However, the figures published in this study tend to undermine this picture of Japan and indicate a far less sharp contrast to the U.S. In response to the question, "Do you think we should reveal the diagnosis of cancer to patients who have requested it?", 69.2% answered "yes," 12.7% indicated they were not sure, and only 17.7% answered no. Similarly, to the question, "Would you wish to be told the diagnosis of cancer if you had cancer?", only 13.2% answered "no." This rate of "no" answers appears, from another study, to have been steadily declining in Japan throughout the 1980s (Morioka, 1991, esp. p. 790). We have seen only slightly higher figures of "yes" and slightly lower figures of "no" in similar polls in the U.S.; realistically, the differences now appear to be almost negligible. Moreover, in the Mizushima Study, larger differences appear between Japanese men and Japanese women than have been found between Japanese citizens and American citizens.

Third, the Kai Study, as I will call it, examined physician-patient interactions in Tokyo, Nagano, and Okinawa. The study assessed how well physicians communicate with patients in a terminal care situation, and in particular assessed the accuracy of the physicians' estimation of the attitudes and beliefs of their patients. The study found that physicians rarely *ask* patients about their preferences in the terminal care situation and that physicians correctly *infer* the patient's preferences in only about 50% of the cases (between 42% and 62%, by region). The sex of the patients appears to make no difference in these rates. The investigators specifically note that their study agrees closely in its conclusions with several studies of physician-patient communications and physician predictions conducted in the U.S. (Kai *et al.*, 1993)

These three studies by Hattori, Mizushima, and Kai tend to discredit the received view – at least, it is discredited if you accept my claims about the past and present situations in the U.S.[12] Nonetheless, even while presenting this evidence the investigators in these studies often report their data as if the received view were correct. Hattori and associates, for example, misrepresent the situation in the United States, and this misrepresentation propels them to perpetuate the received view. Consider the claim in the Hattori study that "The concept of informed consent was elaborated in the United States during the first decade of this century, and was based upon the right of the patient to be fully and clearly informed.... However, ... in Japan, ... before the Second World War ... patients fully entrusted doctors with total and final decision-making authority with respect to their own medical treatment" (Hattori, p. 1008).

I do not question this account of Japanese history, but I have already stated my belief that this account of U.S. history is inaccurate. There was no doctrine of informed consent (as that term is now properly understood) or any requirement of full information in the early part of this century in the United States – or even 15 years after World War II (Faden and Beauchamp, chapters 3–6). The conclusion that I draw is that there is far less difference between Japan and the United States prior to the mid-1960s than the received view holds. The changes that occurred in the U.S. after the mid-1960s had to do almost exclusively with the way medicine and many other parts of American culture were deeply affected by issues and concerns in the wider society about individual liberties, civil rights, and social equality, made dramatic by an increasingly powerful and impersonal medical technology and care. These developments had a deep impact on our culture's appreciation of the importance of informed consent, and the new bioethics in the United States was cradled in this environment.

It seems likely that increased legal interest in the right of self-determination and increased philosophical interest in the principle of respect for autonomy and individualism were instances of the new rights orientation introduced by civil rights, women's rights, the consumer movement, and the rights of prisoners and of the mentally ill – which often included health care components and helped reinforce public acceptance of rights applied to health care. Informed consent was swept along with this body of social concerns, which in turn propelled the new bioethics throughout the 1970s.

There may, of course, be *a gap in time* between a recognition of the importance of informed consent in the United States and a similar recognition in Japan. Perhaps Japan in the late 1980s was like the U.S. in the late 1960s, and perhaps Japan is currently undergoing changes that are propelling it in the same directions the U.S. has taken. Rikuo Ninomiya's brief 1978 article in *The Encyclopedia of Bioethics* suggested at an early stage that several forces were at work in post-war Japan creating an environment of bioethics of just this sort. Ninomiya argued that these factors were the depersonalization of physician-patient contact (together with a depersonalized health insurance system), which caused weakened trust in physicians, a new sensitivity to the rights of individuals in Japan, the indeterminacy and incapacity of traditional moral norms, and the onset of medical malpractice suits (Ninomiya, 1978). These conditions are similar to the conditions that propelled bioethics in new directions in the U.S. in the post-war years.[13] On Ninomiya's analysis, the conditions of change are similar in the U.S. and Japan, even though Japan is a bit slower in bringing the changes about.

I think this historical-lag thesis is far more promising than the hypotheses that underlie the received view, and I suspect the lag thesis can be applied to many other parts of Japanese and American societies.[14]

The Situation Today

Finally, I believe that the received view may have even deeper problems than I have thus far suggested. The available information indicates that the majority of physicians in both the U.S. and Japan *still today* regard disclosure as the primary (and perhaps sole) element of "informed consent." That is, they conceive of informed consent as nothing more than explaining to patients the nature of their medical conditions along with a recommended treatment plan. But if physicians regard informed consent as merely the conveyance of information to patients, rather than a process of discussion with and obtaining permission from the patient, then claims that they regularly "obtain consents" from their patients before initiating medical procedures are vague and unreliable.[15]

Every substantial study of informed consent practices in both the United States and Japan published in English has thrown doubt on whether physicians obtain genuine informed consent from patients, despite the clear increases in patterns of disclosure and the use of consent forms. Thus, a mythology may surround the discussion about informed consent: if physicians *control* information in ways suggested by studies in the United States and Japan, we may be obtaining many fewer informed consents in both countries than we should be obtaining.

I am aware that this conclusion leaves a paradox hanging over my analysis. I have argued that legal requirements and professional codes suggest a vigorous doctrine of informed consent in the United States, but now I am pulling back from that thesis by arguing that the implementation of these requirements in practice may be so weak as to render them broadly inefficacious. This is, I believe, the correct view about the U.S., and I suspect there is much to be said for this same thesis in regard to medical practice in Japan. As Kazumasa Hoshino has observed, Japanese physicians are in principle required under the Japanese Criminal Code to obtain informed consent from their patients, but "despite the general post-War trends [favoring informed consent], the Japanese social ethos still permits medical paternalism to go largely unquestioned" (Bioethics in Japan: 1989–1991, 1992, Vol. 2, esp. p. 380). This thesis is too strongly worded to fit the U.S. situation, but it is a thesis that, when suitably weakened, would fit the U.S. as well as Japan.

III. A REJECTION OF RELATIVISM

Thus far I have dealt largely with sociological and historical questions in comparing the U.S. and Japan. Many important *normative* questions also deserve attention. In this section, I address a subject that sweeps beyond America, Europe, and Japan in order to discuss briefly what the role of morality in the narrow sense, and informed consent in particular, *ought to be* internationally. A widespread view, endorsed by a number of figures in bioethics, challenges the validity of moral precepts that are influential and accepted in one region of the world when these precepts are applied to a different culture (Christakis, 1988 and 1993; Christakis, Fox, Faden and Ijsselmuiden, 1992; Durojaiye, 1979; Muller and Desmond, 1992; Ohnuki-Tierney, 1984). They specifically cite American values of research ethics as a representative example of the problem and reject first-person informed consent as an appropriate universal standard. Research protocols designed in technologically advanced countries but applied in developing countries or to persons from such countries are of particular concern. Because they fear "medical-ethical imperialism," these critics defend concepts such as the "culturally relevant" or "culturally sensitive" uses of information, thereby rejecting the trans-cultural applicability of informed consent standards. They note the vast differences regarding the role assigned to autonomous choice, the nature and causes of disease, the nature of research, and forms of permission giving. I will refer to this perspective as *the relativist position*. This relativism allows us to compromise or even void considered moral judgments in order to accommodate the beliefs of local cultures when, for example, carrying out research involving human subjects in those contexts. Other writers have proposed a partial relativism that retains some universalistic components (Barry, 1988; Levine, 1983; CIOMS, 1991, 1993).

The most obvious thesis to urge against the relativist position is the universalistic one I urged at the beginning of this paper: obligations such as those of informed consent express universally applicable moral values that cannot be compromised without compromising morality itself (Angell, 1988). Although I believe this view to be correct, I will not in this section rely on these claims or on my earlier analysis. Instead, I will adapt a strategy used by Carel Ijsselmuiden and Ruth Faden, who argue for the *inapplicability* of relativistic views on factual grounds, without attempting to show their *unjustifiability* on moral grounds (Ijsselmuiden and Faden, 1992).

Relativists who challenge the importation of informed consent from one culture to another question, on three grounds, the appropriateness of using first-person informed-consent standards in nations with no history of their use. They argue (1) that informed consent is culturally insensitive and inappropriate, (2) that potential patients and subjects are of questionable competence or that insurmountable communication problems prevent competent judgments, and (3) that the gravity of clinical interventions or research investigations render informed consent requirements dangerous for certain cultures. I will consider and reject each of these three claims.

(1) *Cultural Inappropriateness*. In many countries, relativists say, persons view their social roles in terms of close relationships, rather than in terms of individual rights or personal ambition. These traditions of close relationships may be present in a large urban region such as Tokyo, or in small villages or tribes, where the head of the household or tribal leader may make decisions. Culture is multiform, and every culture contains values, beliefs, and rituals of overriding importance to its members. These values should not be nullified by values imported from external cultures.

But how adequate are the data on which these claims rest? I tried earlier in this paper to use recent Japanese surveys to challenge various claims that have been made about Japan. In less complex and less thoroughly studied cultures, it is even easier to misrepresent or overstate actual beliefs, social structures, and changing circumstances. For example, how adequate is the still widely held notion that consent on behalf of otherwise competent adults should be obtained from a "trusted village leader" in African societies, because that is the tradition? In many current African societies there is a deep distrust of and ongoing replacement of older traditions of the village leader – who may be little more than another regime's puppet, and may be abusive or corrupt. Similar arguments apply to consent given by heads of households on behalf of women in many societies.

A time-lag or a lacuna often exists in our evidence about many cultures, and before we can claim an unpreparedness for or lack of interest in informed consent in those cultures we have an obligation to obtain solid and recent evidence. Many of these societies are not as tradition-bound or tied to a specific normative culture as relativists have often claimed or assumed. In many cultures major groups exist that have moral and political commitments similar to the commitments that produced the Nuremberg Code and fueled the civil rights movement in the U.S. Thus, although we certainly ought to include a sensitive cultural analysis in the design of clinical research and in policy recommendations, we also need to draw a threshold line of guaranteed international rights (drawn from morality in the narrow sense) that cannot be overturned even by the most culturally sensitive analysis.[16]

(2) *Incompetence*. Second, relativists sometimes point to a lack of competence to consent in many cultures. This proposal is difficult to assess. These claims often rely for their credibility on problems of

resources: it is time consuming and difficult to obtain valid consents from many patients and subjects, especially in developing countries. But limited resources for the obtaining of informed consents are a problem everywhere; they are perhaps the major problem in the U.S. in deterring high-quality consents from patients. However, we should keep these questions about *resources* separate from questions of *competence*.

The idea that normal patients and research subjects in countries other than our own are psychologically incompetent to give an informed consent is worse than false; it is altogether offensive, and likely in the end abusive. It demeans people and fails to respect them in the way those in our own culture are respected. This same shameful attitude on the part of professionals produced widespread support for informed-consent standards in the U.S. in the 1960s and 1970s. We might also bear in mind that a leading judgment of the Nuremberg trials was a requirement of voluntary consent and respect for all persons who do not suffer from a demonstrable deficiency that renders them incompetent.

Even in difficult cases where physicians or investigators do not share a common understanding of health and disease with patients or subjects, no compelling reason exists to conclude that *cultural*, by contrast to *educational*, differences between the West and other parts of the world are so much greater, or of such a different nature, that non-Westerners should be subjected to truncated disclosure and consent practices (Ekunwe and Kessel, 1984). In any event, if the problems are ones of educational level, the proper conclusion would be that different levels of consent practices and rights should be implemented within the same (Western or non-Western) country on the basis of social class, educational status, and the like. To my knowledge, no one has been so bold as to make this proposal. To the contrary, the view has consistently been that additional effort is required to achieve a level of effective communication with those whose backgrounds are different. This argument applies just as well internationally as locally.

(3) *Medical Gravity*. Third, relativists have also argued that the need for interventions or data is so critical in many countries that the time needed to obtain bona fide informed consent should not be taken. The proposal is presumably that weighty cultural needs *generally* preclude obtaining first-person informed consent, not that they are precluded in a few rare cases. Anyone can, of course, concede the point that under emergency circumstances interventions may permissibly occur without consent.

The emergency situation has long been an established exception to requirements of informed consent. Still, I can see no justification for the claim that the gravity of medical, public health, or research interventions *generally* warrants an exception to informed consent. And even if there were a justification, the argument would apply to research and clinical medicine in the U.S. and Europe, just as well as anywhere else.

I conclude that despite the apparent attractiveness of the rule that we should be "culturally sensitive" in the use of moral requirements, no compelling arguments suggest that we can validly dispense with the obligation to obtain first-person informed consent. Another consideration buttresses this conclusion. In modern medicine (and much of biomedical research) it is hard to imagine adequate treatment occurring without adequate information being shared. How is a patient to understand the reason for a recommended treatment without understanding what the treatment is for? If you recommend carotid artery surgery, chemotherapy, or multiple drugs, say, is it not essential that the patient have some understanding of the underlying condition? How otherwise could you secure the needed cooperation of the patient? Moreover, once patients understand the problem and the recommended treatment, they will typically be less fearful if they know the risks and benefits than they would be if they were kept in ignorance of the risks and benefits.

Failures to obtain informed consents in almost all clinical and research contexts are neither a beneficent treatment of patients (a paternalistic concern for their well-being) nor a culturally sensitive compliance with local traditions (a sympathetic concern for the culture). Except in the aberrant case, such acts are simply failures to treat one human being as he or she ought to be treated.

IV. CONCLUSION

In conclusion, I will bring together my early and late themes in this paper. I have contended that moral development and moral invention in cultures create morality in the broad sense but do not change or negate morality in the narrow sense. I have acknowledged that many moral beliefs differ across cultures, but I have linked requirements of informed consent in particular to more general requirements of respect for autonomy that are parts of morality in the narrow sense. In this way I have made informed consent a universal rather than a regional standard.

I have also argued that it is easy to overlook similarities when we probe for cultural differences and that our culturally informed attitudes and histories in the U.S. and Japan do not preclude near complete agreement on questions of informed consent. I have suggested that we seem to be moving toward a more uniform culture of informed consent at the present time in both countries.

This thesis may also be applicable to Japanese law on informed consent, which shows a lag behind, but a definite similarity to American law (see Hattori, p. 1013). All Japan needs is for one celebrated case like the well-known 1989 case of *Makino v. Red Cross Hospital* (29 May 1989) to be decided in the right way. (Makino was only in part, in my judgment, decided in the right way.) The right decision could radically transform thinking about the law and ethics of informed consent – and, in particular, about the latitude of physician discretion in nondisclosure.[17] Latent in the *Makino* case is *exactly* the legal doctrine of informed consent in the U.S., and I suspect that a very similar doctrine will crystallize as the law develops in Japan. The law will then likely have an impact on medicine and medical ethics similar to its massive impact in the U.S. The *Makino* case shows how morally wrong it is to *fail to disclose* critical information and beliefs to patients. Mrs. Makino, a nurse intimately familiar with the Japanese medical system, died in this case, probably because a radically incomplete disclosure was made to her, though one that conformed to many traditional Japanese practices of partial disclosure. However, these themes would involve a range of speculative hypotheses that extend beyond what I have argued in this paper.

Department of Philosophy/Kennedy Institute of Ethics
Georgetown University
Washington, D.C., U.S.A.

NOTES

[1] See, for example, Macklin (1992), esp. p. 245; National Committee on 'The Declaration of Patients' Rights' (1984); Gostin, (1987), which discusses the Kyoto Principles (and rights) developed in 1987.

[2] For an appropriate sense of invention, see Mackie (1977), esp. pp. 30–37, 106–10, 120–24. Mackie does not mean that individuals create personal moral policies, but that "intersubjective standards" are built up over time through communal agreements and decisionmaking. What is morally demanded, enforced, and condemned is not merely a

matter of what we discover in already available basic precepts, but in addition a matter of what we decide by reference to and in the use and development of those precepts.

[3] I will understand "paternalism" to be the intentional limitation of the autonomy of one person by another, where the person who limits autonomy appeals exclusively to grounds of beneficence for the person whose autonomy is limited. The essence of medical paternalism is an overriding of the principle of respect for autonomy on grounds of the principle of beneficence in caring for patients.

[4] For an excellent illustration of an approach that appeals to differences in kind rather than differences in degree, see Kimura (1991), esp. the chart on p. 237.

[5] Examples of the received view are found in Kimura (1986, 1989, 1992, 1988, esp. p. 179–183); Higuchi (1992, esp. pp. 457, 469–70); Lock (1987); Ohnuki-Tierney (1984); Doi, 1985). Yoshimoto Takahshi has suggested that a major reason for the attitudes under discussion is the prevalence in Japan of physician anxiety, fear, and fantasy of omnipotence. See his 'Informing a patient of malignant illness: Commentary from a cross-cultural viewpoint,' *Death Studies* 14 (1990), pp. 83–91, esp. p. 88.

[6] Higuchi rightly notes that "this dichotomy of self-determination versus paternalism is simplistic."

[7] During the 1950s and 1960s, the duty to obtain consent, which had been established in the early years of the twentieth century, evolved into a new, explicit duty to disclose certain forms of information and then to obtain consent. This development needed a new term; and so "informed" was tacked onto "consent," creating the expression "informed consent," in the landmark decision in *Salgo v. Leland Stanford, Jr. University Board of Trustees*, 317 P.2d 170 (1957). The *Salgo* court suggested, without accompanying analysis, that the duty to disclose the *risks and alternatives* of treatment was not a new duty, but a logical extension of the already established duty to disclose the treatment's *nature and consequences*. Nonetheless, *Salgo* clearly introduced new elements into the law. The *Salgo* court was not interested merely in whether a recognizable consent had been given to the proposed procedures. Instead, Salgo latched tenaciously onto the problem of whether the consent had been adequately informed. The court thus created not only the language but the substance of informed consent by invoking the same right of self-determination that had heretofore applied only to a less robust consent requirement. Shortly thereafter, two opinions by the Kansas Supreme Court, in the case of *Natanson v. Kline*, 186 Kan. 393, 350 P.2d 1093, *opinion on denial of motion for rehearing*, and 187 Kan. 186, 354 P.2d 670 (1960), pioneered the use of negligence in informed consent cases, displacing battery.

[8] There were also during this period a few studies of informed consent in human experimentation; see, for example, Park, Covi and Uhlenhuth (1967); Epstein and Lasagna (1969). For changes in the disclosure of a diagnosis of cancer, see Novack *et al.*, (1979); and Oken (1961).

[9] Only blood tests and prescriptions were reported to have proceeded frequently without some type of consent, but about half of the physicians reported obtaining oral consent even in these cases.

[10] See also Lidz *et al.* (1983); Strull, Lo, and Charles (1984); Faden *et al.* (1982); Lidz and Meisel, Vol. 2; Appelbaum and Roth, Vol. 2; Rimer *et al.* (1983). Unfortunately some of this literature tells us nothing about physicians' actual consent practices.

[11] See further, Shirai (1993), a study that shows significant attitudinal differences.

[12] These two studies themselves often seem to hold the received view of the contrast

between American and Japanese society. For example, Hattori and associates say that, "In countries such as the United States and West Germany, informed consent has become an essential component in ensuring that patients become knowledgeable participants in choosing health care options.... In Japan, however, the concept of informed consent has not yet been generally accepted *by the medical profession*. The right of the patient to take part in the decision-making process *to a larger extent* remains ignored" (p. 1007, italics added). However, on a reasonable interpretation of *the medical profession* and *to a larger extent*, those who conducted these studies do not hold the received view as I have presented it.

[13] However, there likely were many other reasons as well. See the influential analysis in Novack *et al.* (1979).

[14] This thesis may offer an explanation for the disparities between American and Japanese attitudes and practices that have been found in some empirical studies. For an instructive comparison using psychiatric patients, see McDonald-Scott, Machizawa, and Satoh (1992).

[15] Matters may be worse than they appear in the Louis Harris poll: perhaps all these physicians understand by "informed consent" is that the patient's signature has been obtained, or perhaps they mean only that some kind of disclosure has been made. I believe the most adequate conception of informed consent is based exclusively on autonomy, and can be expressed as follows: an act is an informed consent if a patient or subject agrees to an intervention based on an understanding of relevant information, the consent is not controlled by influences that engineer the outcome, and the consent given was intended to be a consent and therefore qualified as a permission for an intervention. In short, an informed consent entails that a subject or a patient substantially understands the circumstances, decides in substantial absence of control by others, and intentionally authorizes a professional to proceed with a medical or research intervention.

[16] A plausible set of both culturally sensitive and universal judgments applicable in Japanese culture is suggested in Mizushima *et al.*, p. 154. These recommendations are entirely consistent with my arguments about the universal requirement of informed consent.

[17] The opinion in *Makino* (Nagoya District Court Judgment, May 29, 1989, 1325 Janji 103) is criticized, appropriately in my judgment, in Higuchi, pp. 458ff and Annas and Miller, pp. 373–75. However, it should be noted that the court found in this case that patients have a right to self-determination, although a right that is subject to the physician's discretion about the adverse impact of medical information. For the potential import of the case in international bioethics, see Swinbanks (1989), and Brahams (1989).

BIBLIOGRAPHY

Alfidi, R.J.: 1971, 'Informed consent: A study of patient reaction', *Journal of the American Medical Association* **216** (May), 1325–1329.

Angell, M.: 1988, 'Ethical imperialism? Ethics in international collaborative clinical research' (editorial), *New England Journal of Medicine* **319**, 1081–1083.

Annas, G.J. and Miller, F.H.: 1994, 'The empire of death: How culture and economics affect informed consent in the U.S., the U.K., and Japan', *American Journal of Law & Medicine* **20**, 357–394.

Appelbaum, P.S. and Roth, L.H.: 1982, 'Treatment refusal in medical hospitals', in President's Commission, *Making Health Care Decisions*, Vol. 2, pp. 411–477.

Bai, K., Shirai, Y. and Ishii, M.: 1987, 'In Japan, consensus has limits', *Hastings Center Report* **17** (June), S18–S20.

Barry, M.: 1988, 'Ethical considerations of human investigations in developing countries: The AIDS dilemma', *New England Journal of Medicine* **319**, 1083–1086.

Brahams, D.: 1989, 'Right to know in Japan', *The Lancet* **2**(8655), July 15, 173.

Christakis, N.A.: 1988, 'The ethical design of an aids vaccine trial in Africa', *Hastings Center Report* **8** (June/July), 31–37EM.

Christakis, N.A., Fox, R.C., Faden, R.R., Ijsselmuiden, C.B., *et al.*: 1992, 'Informed consent in Africa' [letters and replies], *New England Journal of Medicine* **327** (October 8), 1101–1102.

Christakis, N.A.: 1993, 'Ethics are local: Engaging cross-cultural variation in the ethics for clinical research', *Social Science and Medicine* **35**, 1079–1091.

Council for International Organizations of Medical Sciences (CIOMS): 1991, 1993, *International Guidelines for Ethical Review of Epidemiological Studies* and *International Ethical Guidelines for Biomedical Research*, CIOMS, Geneva.

Craemer, W.D.: 1988, 'A cross-cultural perspective on personhood', *Milbank Memorial Fund Quarterly* **61**, 19–34.

Doi, T.: 1973, *The Anatomy of Dependence*, New York.

Durojaiye, O.A.: 1979, 'Ethics of cross-cultural research viewed from third world perspective', *International Journal of Psychiatry* **14**, 137–141.

Ekunwe, E.O. and Kessel, R.: 1984, 'Informed consent in the developing world', *Hastings Center Report* **14**, 22–24.

Engelhardt, H.T.: 1986, *The Foundations of Bioethics*, Oxford University Press, New York, pp. 39–103, 366–379.

Engelhardt, H.T. and Wildes, K.W.: 1994, 'The four principles of health care ethics and post-modernity: Why a libertarian interpretation is unavoidable', in R. Gillon and A. Lloyd (eds.), *Principles of Health Care Ethics*, John Wylie & Sons, London, pp. 135–147.

Epstein, L.C. and Lasagna, L.: 1969, 'Obtaining informed consent: Form and substance', *Archives of Internal Medicine* **123**, 682–688.

Faden, R.R. and Beauchamp, T.L.: 1986, *A History and Theory of Informed Consent*, Oxford University Press, New York.

Faden, R. R. *et al.*: 1982, 'A survey to evaluate parental consent as public policy for neonatal screening', *American Journal of Public Health* **72**, 1347–1351.

Feldman, E.: 1985, 'Medical ethics the Japanese way', *Hastings Center Report* **15** (October), 21–24.

Fellner, C.H. and Marshall, J.R.: 1970, 'The myth of informed consent', *American Journal of Psychiatry* **126**, 1245–1250

Gostin, L.: 1987, 'Human rights in mental health: A proposal for five international standards based upon the Japanese experience', *International Journal of Law and Psychiatry* **10**, 353–368.

Hagman, D.G.: 1970, 'The medical patient's right to know: Report on a medical-legal-ethical, empirical study', *U.C.L.A. Law Review* **17**, 758–816.

Hattori, H., Salzberg, S.M., Kiang, W.P., Fujimiya, T., Tejima, Y. and Furuno, J.: 1991, 'The patient's right to information in Japan – Legal rules and doctor's opinions', *Social Science and Medicine* **32**, 1007–1016.

Hershey, N. and Bushkoff, S.H.: 1969, *Informed Consent Study*, Aspen Systems Corporation, Pittsburgh.

Higuchi, N.: 1992, 'The patient's right to know of a cancer diagnosis: A comparison of Japanese paternalism and American self-determination', *Washburn Law Journal* **31**, 455–473.

Hoshino, K.: 1992, 'Bioethics in Japan: 1989–1991', in A. B. Lustig (ed.), *Bioethics Yearbook*, Kluwer Academic Publishers, Dordrecht, Vol. 2, pp. 379–387.

Hume, David: 1772, 'A dialogue', published with *An Enquiry Concerning the Principles of Morals*, Millar, London.

Ijsselmuiden, C.B. and Faden, R.R.: 1992, 'Research and informed consent in Africa – Another look', *New England Journal of Medicine* **326** (March 19), 830–834.

Kai, I. *et al.*: 1993, 'Communication between patients and physicians about terminal care: A survey in Japan', *Social Science and Medicine* **36**, 1151–1159.

Kimura, R.: 1988, 'Bioethical and socio-legal aspects of the elderly in Japan', *Law in East and West*, Waseda University Press, Tokyo, pp. 175–200.

Kimura, R.: 1986, 'In Japan, parents participate but doctors decide', *Hastings Center Report* **16** (August), 22–23.

Kimura, R.: 1989, 'Ethics committees for "high tech" innovations in Japan', *The Journal of Medicine and Philosophy* **14**, 457–464.

Kimura, R.: 1992, 'Conflict and harmony in Japanese medicine: A challenge to traditional culture in neonatal care', in E. Pellegrino *et al.* (eds.), *Transcultural Dimensions in Medical Ethics*, University Publishing Group, Frederick, MD, pp. 145–153.

Kimura, R.: 1991, 'Fiduciary relationships and the medical profession: A Japanese point of view', in E. Pellegrino, R. Veatch and J. Langan (eds.), *Ethics, Trust, and the Professions: Philosophical and Cultural Aspects*, Georgetown University Press, Washington: pp. 235–245.

Levine, R.J., 1991, 'Informed consent: Some challenges to the universal validity of the Western model', *Law, Medicine and Health Care* **19** (Fall-Winter), 207–213.

Lidz, C.W. and Meisel, A.: 1982, 'Informed consent and the structure of medical care, in President's Commission', *Making Health Care Decisions* **2**, 317–410.

Lidz, C. W. *et. al.*: 1984, *Informed Consent: A Study of Decisionmaking in Psychiatry*, The Guilford Press, New York.

Lock, M.: 1987, 'Protests of a good wife and wise mother: Medicalization of distress in Japan', in E. Norbeck and M. Lock (eds.), *Health, Illness, and Medical Care in Japan: Cultural and Social Dimensions*, University of Hawaii Press, Honolulu, pp. 130–157.

Mackie, J.L.: 1977, *Ethics: Inventing Right and Wrong*, Penguin Books, New York.

Macklin, R.: 1992, 'Universality of the Nuremberg code', in G.J. Annas and M. Grodin (eds.), *The Nazi Doctors and the Nuremberg Code*, Oxford University Press, New York, pp. 240–257.

McDonald-Scott, P., Machizawa, S. and Satoh, H.: 1992, 'Diagnostic disclosure: A tale in two cultures', in *Psychological Medicine* **22**, 147–157.

Meisel, A. and Roth, L.H.: 1983, 'Toward an informed discussion on informed consent: A review and critique of the empirical studies', *Arizona Law Review* **25**, 265–346.
Mizushima, Y. *et al.*: 1990, 'A survey regarding the disclosure of the diagnosis of cancer in toyama prefecture, Japan', *Japanese Journal of Medicine* **29**, 146–155.
Morioka, Y.: 1991, 'Informed consent and truth telling to cancer patients', *Gastroenterologia Japonica* **26**, 789–792.
Muller, J.H. and Desmond, B.: 1992, 'Cross-cultural medicine, a decade later: Ethical dilemmas in a cross-cultural context: A Chinese example', *Western Journal of Medicine* **157** (September), 323–327.
National Committee on 'The Declaration of Patients' Rights': 1984, *Kanja No Kenri Sengen (The Declaration of Patients' Rights)*, Tokyo.
Ninomiya, R.: 1978, 'Contemporary Japan: Medical ethics and legal medicine', in W. Reich (ed.), *Encyclopedia of Bioethics*, Free Press, New York, Vol. 3, pp. 926–930.
Novack, D.H. *et al.*: 1979, 'Changes in physicians' attitudes toward telling the cancer patient', *Journal of the American Medical Association* **241** (March 2), 897–900.
Ohnuki-Tierney, E.: 1984, 'My very own illness', in *Illness and Culture in Contemporary Japan: An Anthropological View*, Cambridge University Press, Cambridge, pp. 51–74.
Oken, D.: 1961, 'What to tell cancer patients: A study of medical attitudes', *Journal of the American Medical Association* **175**, 1120–1128.
Park, L.C., Covi, L. and Uhlenhuth, E.H.: 1967, 'Effects of informed consent on research patients and study results', *Journal of Nervous and Mental Disorders* **145**, 349–357.
President's Commission for the Study of Ethical Problems in Medicine and Biomedical and Behavioral Research: 1982, *Making Health Care Decisions*, Vols. 1 and 2, Government Printing Office, Washington.
Richardson, H.S.: 1990, 'Specifying norms as a way to resolve concrete ethical problems', *Philosophy and Public Affairs* **19** (Fall), 279–310.
Rimer, B. *et. al.*: 1983, 'Informed consent: A crucial step in cancer patient education', *Health Education Quarterly* **10**, 30–42.
Shirai, Y.: 1993, 'Japanese attitudes toward assisted procreation', *Journal of Law, Medicine, and Ethics* **21**, 43–53.
Siminoff, L.A. and Fetting, J.H.: 1991, 'Factors affecting treatment decisions for a life-threatening illness: The case of medical treatment for breast cancer', *Social Science and Medicine* **32**, 813–818.
Strull, W.M., Lo, B. and Charles, G.: 1984, 'Do patients want to participate in medical decision making?', *Journal of American Medical Association* **252**, 2990–2994.
Swinbanks, D.: 1989, 'Japanese doctors keep quiet', *Nature* **339** (June), 409.
White, W.D.: 1983, 'Informed consent: Ambiguity in theory and practice', *Journal of Health Politics, Policy and Law* **8**, 99.
Wildes, K. Wm., S.J.: 'The priesthood of bioethics and the return of casuistry', *Journal of Medicine and Philosophy* **18**, 33–49.

JOY HINSON PENTICUFF

NURSING PERSPECTIVES IN BIOETHICS

I. INTRODUCTION

Nursing perspectives in bioethics – nursing's views of what it means to do good for those who are ill – are derived from the collective narratives of nurses' experiences of caring for patients in countless journeys through illness, recovery, and death. For most nurses, the ethical obligations inherent in each journey must be fulfilled within the boundaries of professional interdependence and complex organizational structures.

All of the activities of nursing are focused ultimately on the good done in the interchange between individual nurse and individual patient. Therefore, nursing's ethical perspectives are close-up, attending to "the concrete particularity and detail of reality as it is experienced at the level of our everyday consciousness" (Gardiner, 1969, p. 4).

II. THE NATURE OF NURSE-PATIENT INTERACTIONS

Caring Within Arm's Length

Nursing's perspectives on doing what is good for the patient originate from the unique viewpoint of two persons within arm's length of each other. One person is a nurse, the other a patient. They are within arm's length because human touch and astute, close observation of patient responses are essential aspects of nursing care. What do nurses derive from this perspective? All that can be perceived by the senses may be taken in. We apprehend the unspoken communication of the body, the tense, relaxed, or flaccid state of the muscles; the stance of the two bodies in relation to each other, the stiffness and guardedness that signify fear or pain. Each person is confronted by emotions of the other, emotions that may be subtle and covered, or searingly intense and exposed. Within arm's length, the nurse opens herself or himself intentionally to the situation of the patient, experiencing the patient as a whole person, with

Kazumasa Hoshino (ed.), Japanese and Western Bioethics, 49–60.

needs for comfort, encouragement, and care that are just as important as needs for restoration of physiologic homeostasis.

Nursing's perspectives in bioethics seek to encompass the entirety of human experiencing – how it is for the patient who is ill, and how it is for the nurse committed to doing good for that patient. These perspectives evolve from nursing's rich history of collective narratives – stories of compassion, duty, giving, touching, and close contact with an infinite array of human struggles and transcendence.

The ethical goals of nursing relate directly to human needs, and these goals are either met or failed within nurse-patient transactions. Nursing perspectives therefore are practical, and include all the thoughts, feelings, and actions that go into caring and being responsible for those who need nursing.

Nursing perspectives in bioethics come out of nurses' iterative experiences in day to day nursing care of patients: the initial greeting and evaluation of patient needs, the mundane but intimate actions of cleansing, feeding, making comfortable; administering medicines; explaining procedures; starting IVs, inserting catheters, applying salves, changing dressings; instructing; answering simple questions and answering questions that go to the core of how the patient will live the rest of his life. All of these nursing actions have a notion of good embedded in them. Benner (1991) refers to this skilled know-how of relating to others in ways that are respectful and supportive of their concerns as *ethical comportment*. "Comportment refers to more than just words, intents, beliefs, or values; it encompasses nuances such as stance, touch, orientation – thoughts and feelings fused with physical presence and action" (Benner, 1991, p. 2).

Nursing perspectives in bioethics come out of what it is to be a good nurse, embracing what it feels like, what one must give from within oneself, the disillusionment felt in one's own and others' failures, what one must know, what is learned in encounters with the heroic saving of life, with cures thought impossible, with unexpected deaths, with suffering, and with prolonged dying. The ethical nurse ponders these and brings the sensitivities that are gained in each episode into the next patient encounter. Ethical practice thus evolves as the nurse through experience and reflection comes to understand how to be with patients in ways that allow a valid carrying out of what is good for that patient, what affirms that patient's dignity, what is fair in that patient's situation.

Cultural Traditions and Nursing Perspectives

The usual cultural norms of each society are the substrate for ethical comportment in nursing practice. Schunior (1989) notes

> [o]ur world view is a cultural pattern that shapes our mind from birth . . . We are shaped by this web; it determines the way we think, the way we see what we see. It is our pattern of representation and our response sustains the pattern. Yet any world view is arbitrary to an indeterminable extent. This arbitrariness is difficult to recognize since our world to view is determined by our world view (p. 9).

Our different customs of interactions impress upon us a view that there is little common across cultures, but this view is of the outward, visible forms through which members of the society relate to each other. The meanings that members of a culture and a society create through their modes of relating are not revealed to us by our mere looking. We must talk with them, live among them to apprehend the true nature of their culture, and thereby recognize any commonalities that might exist between our moral precepts and theirs.

The *manner of relating* is unique for each culture, but the fact that in all cultures there is a human *need for relating* provides a portion of shared value and belief in nursing bioethics perspectives. For example, in the United States, the initial introduction of nurse and patient often is quite informal, with each using first names, whereas in Europe and Japan these initial encounters proceed more formally, and use of first names may be regarded as instances of disrespect. I suggest that the manner in which the moral ideals of nursing are carried out in nurses' actions differ according to the standard forms of human exchange unique to each society, but the moral ideals themselves are part of nursing's common perspective: the protection and enhancement of human dignity, the alleviation of vulnerability, the promotion of growth and health, and the enhancement of coping and comfort in the face of hardship.

III. NURSING PERSPECTIVES ON BIOETHICAL THEORY

Nursing perspectives take into account the theoretical conceptions of moral philosophy and bioethics: traditional principles of beneficence, respect for autonomy, and justice; but nursing's ethical precepts must be true to what nurses experience in doing good for patients. The abstract principles must be borne out through human actions in each particular situation according to a reality that is unique for each patient. Nursing ethical perspectives, then, must include all that goes into understanding what makes up the particular good for this patient, what actions affirm the dignity of this person, what actions constitute fair treatment for this individual who needs nursing care. Understanding these things allows nurses to recognize what they must do to fulfill their ethical obligations to patients. This view of ethical practice admits within bioethical theory a new realm of subjectivity and affect that points to rejection of a purely rational and impartial ethic. Likewise, nursing's modes of seeking understanding are diverse, drawn not only from the social and natural sciences, but also encompassing our intuitive, aesthetic understanding of the harmonies and discordancies of each patient's life. Nursing's traditional modes of understanding have long embraced an entering into experience with the patient, with a focus on the narrative of each patient's life that allows an intelligible answer to the question, 'What is good for this patient?' but our recent elevation of the scientific as the paramount mode of understanding threatens to distort nursing's humanistic perspective.

The affect-neutral, rational, impartial approach inherent in science cannot supply the intellectual and affective resources necessary for ethical understanding in nursing practice. As Benner (1990) notes, the moral dimensions of caring in nursing require attention to the local and the specific, the particular and the concrete, not just the abstract and the theoretical. This same inadequacy occurs when an impartial approach dominates bioethics orientations. Impartial, distant, affect-less approaches in both science and bioethics are often antithetical to the caring perspective that has emerged as a significant aspect of nursing ethics within the past decade (Huggins and Scalzi, 1988; Omery, 1989).

Nursing's reaffirmation of its commitment to caring (Benner, 1988; 1990; Leininger, 1988; Watson, 1985; Fry, 1989a; 1989b; Valentine, 1989) is an important nursing perspective in bioethics. Carper (1979) notes that caring requires an understanding of the person as a whole being. She states, "To be concerned with the 'whole person' and to practice with consideration and sensitivity for the integrity of the human self is basically an ethical injunction" (pp. 11–12). Benner (1990) argues that "a biomedical ethics based on problem-oriented issues of iatrogenesis and questions of autonomy, rights, justice and paternalism ... cannot provide a positive statement of the good and must be augmented by an ethic of care" (p. 1).

Caring requires connectedness, context-bound values and strategies, conceptions of the self as embedded in relationships, and an emphasis on the good within communities. In caring, human relationships are thought of in terms of sensitivity and responsiveness to needs. Within caring perspectives, moral conflict, conflict resolution, and understanding of good and harm take on new definitions (Gilligan, 1987).

IV. CARING WITHIN INTERDEPENDENT ROLES AND ORGANIZATIONAL STRUCTURES

A second nursing perspective comes from our recognition that no one discipline has all the resources necessary to sustain the patient in his journey through the valleys and shadows of the human condition. Each discipline has its own special gifts that are needed, and each has some resources common to all the helping professions. But for those special talents that nursing practice traditions have tended and enriched, our discipline assumes special ethical obligations of stewardship. I see this notion of special gifts and special obligations as true for all the helping professions. For example, if we take it to be a moral ideal that nursing seeks to alleviate vulnerability, then the enacting of that ideal in practice carries special obligations that nurses be open to the patient's suffering and competent in using nursing's special gifts to assuage it.

Baker and Diekelmann (1994) note that because as nurses we believe that basic care starts with listening carefully for the emerging story of patients' and families' lived experiences, we are uniquely situated to attend to the meanings they form around the experience of being ill, of recovering, or of living toward death. In contrast, because physicians' gifts involve attention to the cure of disease rather than to the experiencing of it, medicine's perspectives are derived from different activities in the employment of medicine's unique talents essential to patients' welfare. Thus, the helping disciplines share many moral ideals, but the unique resources and talents of each discipline give rise to correspondingly unique ethical obligations that add distinctive content to each discipline's perspectives in bioethics.

Bioethics Within Constraining Contexts and Multiple Obligations

In the United States, some voices in nursing say that the ethical ideals in nursing care cannot be realized in our current economically and technically driven systems of health care delivery (Hall and Allan, 1994; Benner, 1991; Baker and Diekelmann, 1994). Studies of nurse burnout (Jacobson, 1978; American Hospital Association, 1987; Bartz and Maloney, 1986; Cameron, 1986; Cronin-Stubbs and Rooks, 1985; Keane, Ducett and Adler, 1985) reveal that disillusionment with the work of nursing, apathy in the face of patient need, moral distress and moral outrage are frequent aspects of nurses' experience. Hall and Allen (1994) note that "many of the environments in which nurses practice are sterile and noncaring for everyone" (p. 115).

How caring is maintained or eroded is significant to understanding nursing ethical practice because if nurses do not have the will to practice ethically, the entire enterprise of nursing as a helping profession is in jeopardy. The demands upon nurses to provide competent, compassionate care can result in moral distress if nurses are unable to meet their own or others' expectations. Moral distress is defined by Wilkinson (1986/87) as "the psychological disequilibrium and negative feeling state experienced when a person makes a moral decision but does not follow through by performing the moral behavior indicated by that decision" (p. 16). Wilkinson concluded, "Those nurses who are unable to cope with moral distress and who leave bedside nursing seem to be those who are most aware of, and sensitive to, moral issues, and who feel a strong sense of responsibility to patients and for their own actions" (1986/87, p. 27).

In the United States, Prescott and Dennis' (1985) study of power and powerlessness in hospital nursing departments and Alexander, Weisman, and Chase's (1982) research on determinants of staff nurses' perceptions of autonomy within clinical contexts conclude that nurses' moral agency is impeded by institutional limitations of their power and by the divided loyalties of the nurse to patients, nursing peers, physicians, and institutions. Nursing practice was found to be fraught with conflicts between nurses' personal values and the requirements of administrative policy. These studies conclude that in the United States nurses do not have the autonomy and organizational influence necessary to carry out their professional responsibilities for patients (Institute of Medicine, 1982; American Academy of Nursing, 1983; Dennis, 1983; Prescott and Dennis, 1985).

Numerous questions surround this issue. Do practice context considerations modify nurses' normative ethical obligations? What are nurses' options as moral agents within the health care settings? How adequate are institutional resources such as ethics committees for resolving ethical dilemmas that nurses face in practice?

V. CARING COMMUNITIES

Many nurses in the United States are calling for the transformation of health care systems into caring communities – places and systems in which the good for patients can be instantiated and worked out. Hall and Allan (1994) depict caring communities as relationships in which all support and define the importance of giving in interrelationships. Ray (1994) notes, "... a moral community may be viewed as the experience of shared meanings ... where moral virtues, values, and principles guide inter-relational behavior toward responsible choice making for the good of the whole" (p. 107). Speaking as a nurse involved in the current United States debates of health care reform, I might add that the primacy of individual freedom and the view of health care as a market commodity so characteristic of our society makes the transformation of our current health care system into caring communities implausible. I believe that the importance of the common good acknowledged in the bioethical perspectives of almost all countries other than the United States is a basic starting point for health care systems that can instantiate notions of caring and ethical comportment in practice.

VI. COMMONALITIES AND CONTRASTS IN INTERNATIONAL NURSING BIOETHICS PERSPECTIVES

Commonalities

Whether nursing perspectives in bioethics share important commonalities across cultures and societies remains open to debate, but if those in the helping professions can be successful in creating communities of caring, I suggest that there are universals to be found. If we start from a commitment to do good for others, and if bioethics is seen as inherent in practice rather than as called for only in quandaries, then we might be open to the view that the ways we seem to differ as human beings are less significant to ethical practice than the universality of human needs and human compassion.

Because of communication technology, our knowledge of ethical issues faced by people in other parts of the world and how these issues are dealt with has increased remarkably. The issues broadcast around the globe – dramatic news – typically are in the form of intense dilemmas. We may sense a fundamental doubt that there could be any commonalities between our ethics and those shaped by cultures and societies so different from our own. Recently in the United States, for example, there was much dismay expressed at the public caning in a foreign country of a young American man who had defaced others' property. Some Americans found this form of punishment demeaning and abhorrent. It is interesting to note that recently, however, the city of San Antonio, Texas debated the use of public spanking to punish those who deface public buildings.

I propose that when nursing's view of bioethics is from the perspective of everyday ethical comportment pervasive in practice, rather than from the perspective of ethics as conceptual tools to solve quandaries, there are many more commonalities than differences in nursing ethical perspectives across the globe. Certainly there is great range in the scope of nursing practice in Europe, Israel, South East Asia, Canada, Australia, and the United States; but it seems reasonable to assume that nurses in each of these countries believe that when they entered the profession of nursing they accepted an obligation to help others. Given this starting point, I would suggest that our common vulnerabilities as embodied selves to illness, loss, and death, and our common realization of the possibilities of transcending these vulnerabilities provide enough of what is needed to construct a common foundation for nursing ethics and for bioethics generally. That said, I now turn to an area in which I believe nursing perspectives show striking contrasts from country to country.

Contrasts

One particularly forceful influence on nursing perspectives in bioethics is the extent of medical technology and intensive care available in countries, and the expectations of societies that technologically advanced therapies will be employed and will be successful in rescuing lives from death. I believe that differences in nursing perspectives often reflect the influence of technologic advance, because technologic approaches alter a society's views of science and the goals of medicine. Technology acts upon persons, using a rational science of prediction and control that may be antithetical to traditional cultural beliefs about the nature of persons, disease, healing, and death.

In the United States, medical advances have been accomplished primarily through increased understanding of pathophysiology of diseases and pharmacologic and surgical interventions to correct or remediate pathological processes. We export this medical technology and our bioengineering industry's production of sophisticated diagnostic and life support equipment makes up a significant portion of the gross domestic product. The widespread use of sophisticated technologies of diagnosis and life support is characteristic of health care in the United States. As a society, we have been seduced by this technology into believing that we do not have to be responsible for the maintenance of our own health because our lives can be restored through our technology. We did not worry about what the technology cost; we did not question whether the technology produced more good than harm. The consequences of our unquestioning embrace of medical technologic advance has resulted in erosion of the influence of humanistic approaches in American health care.

Nursing ethical perspectives have been especially challenged during this period of biotechnical dominance, because the technologic approach so often devalues the need for nurses and patients to be co-adventurers in the patient's journey. Nursing's view of patients as whole people whose illness has compromised that wholeness is shattered by an approach that requires that human beings be thought of as composites of organ systems. Most medical textbooks, and until recently, most nursing textbooks have been organized according to cardiopulmonary, renal, digestive, or nervous system dysfunction. This conceptualization applied to persons who are ill is radically misleading, because it omits all except the physiology of illness. In the United States we have taught physicians and nurses how to take care of bodies, but we have not taught them how to take care of persons who are ill and suffering. When we export the equipment that provides the means for this technologic approach to illness, we should warn those who buy it that it carries the potential to change fundamentally their society's views of what it means to do good in the allocation of scarce resources and what it means to sacrifice other values to the primacy of saving physiologic life.

VII. CONCLUSIONS

The nursing ethical ideal of caring, engagement, and use of the nurse's whole self in nurse-patient relationships so as to know what the good is and how to bring it about for particular patients is being affirmed in international nursing literature, conferences, and conversations among nurses in many countries. These ethical ideals are congruent with cultural valuing of the common good and traditions that foster understanding of persons as whole beings. I close with some thoughts from the previously mentioned work of Schunior (1989) about nursing perspectives on balancing a technology of predicting, controlling, and intervening – what Schunoir calls the wearing of the tragic mask – with the human responses of touching, apprehending, and feeling – the wearing of the comic mask. He writes,

> Curing . . . is a mechanical, functional restoration. But because fatality, loss, and change are the inescapable substrates of our existence, an exclusive emphasis on cure is necessarily frustrated and tragic. Healing also involves restoration, but in an existential rather than functional mode, a restoration to the center of one's being (Schunior, 1989, p. 15).

Schunior refers to Carse's description of healing through "touching" as opposed to "moving" persons.

> You move me by pressing me from without toward a place you have already foreseen and perhaps prepared. It is a staged action that succeeds only if in moving me you remain unmoved yourself (Schunior, 1989, p. 15).

The nurse of the comic mask relates to patients without the expectation or desire of controlling or mastering experience but only to enter into experience. And because we know there is no touching without being touched, we are open and vulnerable with our patients. Accepting patients as they are, where they are, without the tragic hero's concern with defining, controlling, and changing, permits the spirit to expand and express itself (Schunior, 1989, p. 16).

Nursing's bioethical perspectives, self-definition, and emerging willingness to enter into the patient's reality are producing an increasingly truthful understanding of what good is for the patient and what our role may be as co-adventurers toward that potential good. We are in the midst of epistemologic change in bioethical perspectives, a change that I believe in the next decade or so will be influenced significantly by nursing's affirmation of feeling, apprehending, touching, and putting aside the tragic mask.

School of Nursing
University of Texas at Austin
Austin, Texas, U.S.A.

BIBLIOGRAPHY

Alexander, C.S., Weisman, C.S. and Chase, G.A.: 1982, 'Determinants of staff nurses' perceptions of autonomy within different clinical contexts', *Nursing Research* **31**(1), 48–52.

American Academy of Nursing: 1983, 'Magnet hospitals: Attraction and retention of professional nurses', American Nurses Association, Kansas City.

American Hospital Association: 1987, 'Moral distress in nursing', *Hospital Ethics* **3**(4), 1–4.

Baker, C. and Diekelmann, N.: 1994, 'Connecting conversations of caring: Recalling the narrative to clinical practice', *Nursing Outlook* **42**(2), 65–70.

Bartz, C. and Maloney, J.P.: 1986, 'Burnout among intensive care nurses', *Research in Nursing and Health* **9**(2), 147–153.

Benner, P.: 1991, 'The role of experience, narrative, and community in skilled ethical comportment', *Advances in Nursing Science* **14**(2), 1–21.

Benner, P.: 1988, *The Primacy of Caring*, Addison-Wesley, Menlo Park.

Benner, P.: 1990, 'The moral dimensions of caring', in J.S. Stevenson and T. Tripp-Reimer (eds.), *Knowledge About Care and Caring*, American Academy of Nursing, Kansas City, pp. 5–17.

Bartz, C. and Maloney, J.P.: 1986, 'Burnout among intensive care nurses', *Research in Nursing and Health* **9**(2), 147–153.

Cameron, M.: 1986, 'The moral and ethical component of nurse burn-out', *Critical Care Management Edition* **17**(4), 42B–42E.

Carper, B.A.: 1979, 'The ethics of caring', *Advances in Nursing Science* **1**(3), 11–19.

Cronin-Stubbs, D. & Rooks, C.A.: 1985, 'The stress, social support, and burnout of critical care nurses: The results of research', *Heart and Lung* **14**(1), 31–39.

Dennis, K.E.: 1983, 'Nursing's power in the organization: What research has shown', *Nursing Administration Quarterly* **8**, 47–57.

Fry, S.T.: 1989a, 'Toward a theory of nursing ethics', *Advances in Nursing Science* **11**(4), 9–22.

Fry, S.T.: 1989b, 'The role of caring in a theory of nursing ethics', *Hypatia* **4**(2), 88–103.

Gardiner, P. (ed.): 1969, *Nineteenth-Century Philosophy*, The Free Press, New York.

Gilligan, C.: 1987, 'Gender difference and morality: The empirical base', in Kittay, E.F. and Meyers, D.T. (eds.), *Women and Moral Theory*, Rowman & Littlefield, Totowa, N.J.

Hall, B.A. and Allan, J.D.: 1994, 'Self in relation: A prolegomenon for holistic nursing', *Nursing Outlook* **42**(3), 110–116.

Huggins, E.A. and Scalzi, C.C.: 1988, 'Limitations and alternatives: Ethical practice theory in nursing', *Advances in Nursing Science* **10**(4), 43–47.

Institute of Medicine: 1982, 'The study of nursing and nursing education', National Academy of Medicine, Washington, D.C.

Jacobson, S.F.: 1978, 'Stressful situations for neonatal intensive care nurses', *The American Journal of Maternal Child Nursing* **3** (May/June), 144–150.

Keane, A., Ducette, J. and Adler, D.C.: 1985, 'Stress in ICU and non-ICU nurses', *Nursing Research* **34**(4), 231–236.

Leininger, M.: 1988, *Care: Discovery and Uses in Clinical and Community Nursing*, Wayne State University Press, Detroit.

Omery, A.: 1989, 'Values, moral reasoning, and ethics', *Nursing Clinics of North America* **24**(2), 499–508.

Prescott, P.A. and Dennis, K.E.: 1985, 'Power and powerlessness in hospital nursing departments', *Journal of Professional Nursing* **1**(6), 348–355.

Ray, M.A.: 1994, 'Communal moral experience as the starting point for research in health care ethics', *Nursing Outlook* **42**(3), 104–109.

Schunior, C.: 1989, 'Nursing and the comic mask', *Holistic Nursing Practice* **3**(3), 7–17.

Valentine, K.: 1989, 'Caring is more than kindness: Modeling its complexities', *Journal of Nursing Administration* **19**(11), 28–34.

Watson, J.: 1985, *Nursing: Human Science and Human Care*, Appleton-Century-Crofts, CT, (1988, 2nd printing) Colorado Associated University Press, Boulder.

Wilkinson, J.M.: 1986/87, 'Moral distress in nursing practice: Experience and effect', *Nursing Forum* **23**(1), 16–29.

EMIKO NAMIHIRA

THE CHARACTERISTICS OF JAPANESE CONCEPTS AND ATTITUDES WITH REGARD TO HUMAN REMAINS

I. THE UNIQUENESS OF THE JAPANESE STYLE OF CREMATION AND A RITUAL OF PICKING UP THE CREMATED BONES

In order to understand Japanese people's concepts and attitudes regarding a dead body we must first know how they dispose of a dead body and why they do it in that particular way. Present day Japanese use cremation in more than 90 percent of deaths. However, it has been only since the Second World War that more than 70 percent of the dead have been cremated. During the Tokugawa period cremation was employed only by the people of the Jodo-shinshu Buddhist sect, and even in this sect cremation did not always take place. The usual method of disposal was burial, except in Kyoto district, which used to be the capital of Japan, with a high population density since the 10th century. During the second half of the nineteenth century cholera epidemics killed almost a million Japanese and after the epidemics the custom of cremation spread widely because of the large number of people dying during such a short period and the new awareness among them of the dangers of infection.

The style of cremation employed in Japan is very different from that of other countries. In fact, no other culture in the world uses such a style. The body is cremated only as far as the bones; it is not reduced into ash. The bones must remain in the shape of the body with each bone in place. This style of cremation requires a highly sophisticated technique, and just one cremation oven built in 1992 cost a hundred million yen. The cremated remains are gently and carefully withdrawn from the furnace so that the bones remain in the shape of the body. After the cremation the family, close relatives and sometimes close friends of the dead person gather around the bones and pick up each piece of bone in turn with chop sticks. This action of picking up bones and putting them into a special jar is a very important ritual in the Japanese funeral ceremony.

Kazumasa Hoshino (ed.), Japanese and Western Bioethics, 61–69.

Not only adults but also young children, and sometimes babies younger than two years, participate in the ritual.

Bones are placed into a ceremonial jar in a fixed order. That is, a piece of foot bone is placed at the bottom of the jar, a piece of leg bone is put above it, and then a piece of thigh bone above that and so on. Finally the skull is set on the top. In some cases above the skull they place a piece of back bone which people believe to be "the Adam's apple," the so-called Japanese "Buddha's figure" bone. (In fact the "Adam's apple" has been burned away, since it is cartilage.) For the Japanese the bones in the jar are the actual body of the dead person. The bones in the jar must not be divided nor put into two or more jars. A division of the bones is prevented by the crematoriums and the Buddhist priest who is conducting the funeral ceremony. The reason given for this taboo is that the dead spirit cannot mature or begin to proceed towards ancestorhood if these bones are divided.

When Japanese people used burial as a way of disposing of a dead body, families usually placed the dead body immediately above its predecessors in the same family graveyard plot. Therefore, they could watch the process of the decomposition of the bodies (i.e.; whether the bodies that had been buried earlier had become white bones or not), every time they buried a new body in the same plot. Furthermore, the Japanese wooden coffin rapidly broke up after the burial because it was fragile and the soil in Japan is often soggy. The shape of a traditional coffin was like a tub rather than the standard flat box of today. A dead body was put into the coffin in a sitting position with crossed legs and folded arms. The fixed order in which people put the cremated bones into a jar is the order in which the dead body in the tub-shaped coffin decomposed over a long period of time.

In the Ryuku and Amami Archipelagoes they dig up the bones six years after death, wash them and rebury them, apart from the skull, which is stored in a beautiful large jar with the skulls of other dead members of the family. They believe that the dead spirit has matured when the bones have become white. People in the mainland islands have not developed such a ritual, but they have other ways of seeing the decay process.

II. A RITUAL OF VIGIL : THE CRITICAL TIME BETWEEN DEATH AND CREMATION

Before cremation Japanese people perform a kind of wake. This ritual is a pre-funeral ceremony, and for the family and the close relatives of a dead person it is more important than the actual funeral. They keep vigil all night and receive visitors who are expected to come any time during the night to view the dead body. The viewing is called "shinime-ni-au," meaning "to see the person dying" or "to be present at the exact moment of death." The cremation is sometimes delayed until a member of the family has returned from a distant place. While traditionally the vigil must always precede the actual funeral, at times in the modern world people die in a hospital remote from home and are cremated without a vigil, or the vigil takes place after cremation. This is very unusual and unfortunate for both the family and the dead person himself. To avoid such circumstances the family sometimes asks the doctor to predict the time of death and they plead with him to move the patient to a hospital near their home. Occasionally, when a person has died suddenly (e.g., on a business trip), the dead body is returned home from a distant place although this is very expensive.

The period between death and the end of the funeral ceremony, especially the cremation of the dead body, is the most confused and critical time both for the dead person and his family. The dead body in this period is thought to be impure and potentially contaminating. It is still essential all over Japan to put a short sword or a mirror on a chest of the dead body. This ritual is to protect the dead body from an "evil" or "unknown" spirit, and to prevent the body from becoming very dangerous to the people who surround it. To avoid this situation a ritual to diminish the impurity is begun as soon as the death occurs, and it is performed all over Japan even when death takes place in a hospital. Each hospital ward has a special kit for cleansing a dead body. In most cases the nurse who cleanses the dead body wears a special gown, and in some wards the nurse also has a face makeup kit. Professional medics are not aware that this cleansing is a ritual, but they are very careful not to damage the body and treat it gently and sensitively. They often use hot water so that "the dead person does not catch cold" or "is not surprised by contact with cold water." This cleansing ritual is still found all over Japan. In some districts all close-kin female members gather and cleanse the whole body. People in urban areas no longer have a clear idea that a dead body may be contaminating the family and the people around it, as the Japanese traditionally understood. However, even such urban people scatter salt to purify their own bodies before entering their house after participating in a vigil and funeral ceremony. In rural areas, where the people still have the idea of death contamination, the bereaved family and the people who assisted at the funeral ceremony purify the house, and all tools and equipment used in the funeral, by washing them with salt and water or scattering salt. The law allows family mourners and close kin of the dead person to be away from work and school. The length of time off depends on the closeness of the relationship, i.e., spouses, parents or children can take a week off and grandchildren and siblings can take a maximum of three days because the degree of contamination depends on the degree of closeness in social relationship not on the degree of physical closeness. The mourners at one time were obliged to stay at home and avoid seeing other people for 49 days after a funeral. This leave is called "kibiki" in Japanese, which literally means "leave for mourning" but actually means "keeping indoors so as not to contaminate other people or public spaces."

III. THE MEMORIAL CEREMONIES AND THE PERSONHOOD OF THE DEAD

Even from the above brief illustrations we can see the contradictory attitudes of Japanese people toward a dead body, especially where the dead person is a family member. That is, on the one hand, they recognize the impurity and contamination caused by the dead, while, on the other hand, they have developed rituals in which the bereaved family has many opportunities to touch the dead body. Contact with the actual bones of the dead continues for 33 years, in some districts for 49 years, i.e., the people visit the family plot where the cremated bones are kept before or after all the memorial ceremonies performed by the offspring for 33 or 49 years. This visit is obligatory. The Japanese style of cremation is unique, and the period of years over which the memorial ceremonies are performed is particularly long. In most Buddhist countries, e.g., Korea, Thailand, and Burma, people perform memorial ceremonies for each dead person for only seven years. I can suggest that the length of time in the Japanese case is related to the particular value that the Japanese family system puts on the perpetuation of every family line.

It is supposed that the custom in which each family performs a long series of the memorial ceremonies for each dead person was adopted by the common people in the first half of the 19th century, and it is also supposed that two factors served in the formation of such customs. One was the institution called "danka-seido" ("terauke-seido") in which all families of a community had to be members of the community temple (temples) and every human death had to be reported to and documented by the temple. In this system the details of each death were documented and Buddhist priests were able to exhort the temple members to perform memorial ceremonies for the dead. Another element was the increase in the number of independent farmer families during the period and the fact that belief in the perpetuation of each family line filtered its way into these independent farmer families. The value of the perpetuation of each family line was strengthened by the Family Law that was enacted in the last decade of the 19th century and lasted until 1947, although the institution of "danka-seido" was abandoned five years after the Meiji Restoration of 1868.

The relationship between dead spirit and the offspring of a family is very reciprocal. If the offspring of a family perform the memorial ceremonies and make sufficient punctual offerings to the dead spirits, the spirits receive the power to bless the family, and the family can get prosperity and the perpetuation of the family line. Then, the prosperity and the perpetuation of the family enable the family member to make offerings.

IV. METAMORPHOSIS OF A DEAD BODY AND CHANGING OF THE IDENTITY

Returning to the question of Japanese people's contradictory attitudes to a dead body, I suggest that such attitudes rise from the critical situation of a dead body between death and cremation and the Japanese concept of "person." When a person dies, his body remains but is no longer alive. It does not answer, smile or react to other people's actions, i.e., no kind of communication can take place between the dead body and other people. However, the body retains the physical characteristics of the person until cremation. Therefore, we understand cremation as a cultural or artificial process to metamorphose a dead body. All cultures have ways of making a human body metamorphose, but they do not leave a human dead body to rot and decompose as the body of an animal in the wild would. The Japanese style of cremation is one of the most sophisticated methods of disposal or artificial metamorphosing of a human body, and the ritual of picking up the cremated bones and putting them into a jar is also one of the most elaborate procedures to represent the change of a human body from a living state to a dead state. Such an elaborate procedure in disposing of a human dead body has been developed in order to acknowledge the process in which a person loses his identity after death. The ritual of viewing the dead person's face and body during the vigil is performed in order to recognize the identity of that person during his life time, and the action of picking up pieces of bone in turn after the cremation acknowledges the change in the body's identity. During the period between death and disposal of the person, the body is in a transitional state: from life to death, from a living body to a dead body, from the world of the living to the world of the dead. Therefore, the period between death and the end of the funeral ceremony is the most confused and critical time both for the dead person and for his family. The dead person is losing his lifetime identity but has not yet acquired a new identity. In this situation both the body and all materials associated with it are considered in many cultures, including the Japanese, to be very impure and potentially contaminating.

Anthropological and ethnological data show that people in many cultures recognize such a transitional state to be impure and dangerous. A woman in childbirth, a new born baby, young boys and girls in initiation ceremonies, the last day of the year, the time between a winter season and a spring season, and the transitional states between life and death are all recognized as impure and dangerous. Japanese culture is one of those that recognize most strongly the impurity of death. A ritual meaning of memorial ceremonies is to remove or diminish the impurity caused by death. In Japan the dead body becomes increasingly contaminated and dangerous just before cremation. To avoid this situation a ritual to reduce impurity must begin immediately after death. The cleansing of a dead body, even within a hospital, is a ritual to diminish the impurity, although professional medics never recognize this fact. Traditionally the spirit of a dead person who died outdoors, or who died without the presence of his family, is thought to be more impure and more dangerous because the cleansing rituals cannot be performed immediately. I believe that nurses' cleansing of a dead body is in accordance with this belief, i.e., they think that if the family cleanses the body after bringing it home, it is too late and that the cleansing ritual must take place at the actual location of death. The growth toward maturity of the dead spirit is also recognized to be the reduction of the impurity of death, and the ancestorhood reached by a dead spirit is believed to be perfectly pure. The Japanese people's contradictory attitudes toward dead bodies, which are revealed in vigil and funeral ceremonies, emphasize the transitional state of the dead person, and show that Japanese people think that the physical element, i.e., the body, is an important part of personhood, and that death itself cannot reduce the personhood of the body.

V. THE JAPANESE CONCEPT OF PERSONHOOD

A British anthropologist, A. R. Radcliffe-Brown, interpreted the concept of "individual" as biological organism and the concept of "person" as a complex of social relationships, but in Japanese thinking these two concepts are as one and cannot be divided. According to his interpretation, death marked the end of the "person," but the Japanese believe that a "person" can exist for many years after death, although his identity partially changes. His physical identity is also retained because the family takes care of the jar which contains his bones, and the bones are believed to be his actual body. In funeral and memorial ceremonies the physical and social identities are brought together.

In her essay 'Person and individual, some anthropological reflections,' J.S. La Fontaine discussed the concepts of "person" and "individual" in four societies. For the Guhuku-Gama people in Highland New Guinea, individual and personal identity are two sides of the same coin. The same can be said of these concepts among the Japanese, although Japanese society is very different from that of the Guhuku-Gama. For the Japanese a body is both material and immaterial, biological and social. The entity of the immaterial is manifested in a physical body and it is the immaterial that enables this physical body to have a social identity. This complex two-sided nature of personal identity is hidden while a person is alive. When he dies, the double-sided nature of personhood becomes obvious to the people who have been in social contact with him. They recognize that the interrelationships between the two sides are revealed at death. People have to handle this critical situation. The ritual of picking up the cremated bones is the manifestation of this recognition.

At the same time the complex rituals of a funeral ceremony are the means by which people reunify the two sides of the "person." During the funeral procedure the identity of the two sides, that is, the physical and the social, the material and the immaterial, are partly changed. It follows that the time period after a person has died, while the two parts of identity have not reunified, is the most unstable and confused for those people who have had a social relationship with the dead person.

VI. CONCLUSION

In the case of a cadaver organ transplant an organ must be removed from the body before those procedures by means of which the identity of the body in life is altered by cremation and the family of the dead openly acknowledge the metamorphosis and decay of the body. As I have suggested above, for Japanese people the cremation is not only the way of disposing of a dead body, but also a way to change the identity of the body from living to dead. In traditional Japanese thinking cremation is a way of changing the life time identity rather than disposing of it. Therefore, for Japanese people, removing an organ from a dead body before such procedures have taken place represents a failure to create a new bodily identity for the dead person. Cremated bones are not seen, by the Japanese, as human remains but as the actual body of the dead person. Since Japanese people think that the dead "person" also consists of the physical and the spiritual, the dead fails to acquire a new "personhood" if his body is not metamorphosed by means of the correct procedures. If the dead fails to achieve a new personhood, or acquires an injured or imperfect personhood, the dead will not able to achieve ancestorhood.

In traditional Japanese thinking the dead spirit can achieve "religious growth" through the memorial ceremonies. After the long series of memorial ceremonies over a period of 33 or 49 years the dead spirit can reach ancestorhood, and thereby bless his offspring. However, this growth cannot be achieved by the dead spirit on its own, but can only take place by means of the family's offerings. Therefore, the relationship between a dead spirit and his offspring is very reciprocal as I mentioned above. It is my suggestion that this reciprocal relationship represents the Japanese idea that "personhood" remains after death.

After the Second World War the Family Law was reformed, and the commonly held belief in family perpetuation lost its legal base. Rapid industrialization during the 1950s and 1960s caused population movements from rural areas to urban areas, and also caused widespread urbanization all over Japan. It has become difficult to maintain the concept of family perpetuation during this process. Funeral and memorial ceremonies are still performed, although the traditional style has been changed. People do not seem to hesitate to spend a large amount of money on a funeral ceremony. I believe, however, that the difficulty of a dead person is deeply embedded in the Japanese idea of personhood, which is based on Japanese social relationships and Japanese concepts of life and death.

Kyushu Institute of Design
Fukuoka, Japan

BIBLIOGRAPHY

La Fontaine, J.S.: 1985, 'Person and individual: Some anthropological reflection', in Michael Carrithers *et al.* (eds.), *The Category of the Person*, Cambridge University Press, Cambridge, pp. 123–140.

Namihira, E.: 1985, *Kegare (The Impurity)*, Tokyo-do, Tokyo, pp. 150–185.

Radcliffe-Brown, A.R.: 1940, *Structure and Function in Primitive Society*, Cohen and West, London.

PART TWO

TRADITIONS, AUTHORITIES AND AMBIGUITIES

GERALD P. MCKENNY

TECHNOLOGY, AUTHORITY AND THE LOSS OF TRADITION: THE ROOTS OF AMERICAN BIOETHICS IN COMPARISON WITH JAPANESE BIOETHICS

Bioethics as a modern field of study is still, relatively speaking, in its youth, and like most youths it is not disposed to reflect on its roots. When bioethicists do discuss the roots of their endeavor, they typically trace the rise of bioethics to the moral quandaries of modern technological medicine or to the crisis of moral authority that challenged the competence or the right of medical and religious traditions to resolve moral issues raised by medical practice. It is, of course, in the interests of standard bioethicists to find the roots of their movement in the quandaries modern technology presents or in the need for a common morality. If technology presents moral problems that standard bioethics alone can resolve or if standard bioethics alone can claim a public moral authority that traditional moral schemes have lost, then standard bioethics would be not only culturally inevitable but also rationally necessary. Moreover, if similar problems raised by technological medicine or a similar crisis of moral authority are present in other cultures then standard bioethics would supply a culture-transcending morality. In our case, this would mean that American bioethics would provide a more or less perfect match for transplantation into Japanese culture.

By contrast, I show how standard bioethics is not a rationally necessary response to technology or to a crisis of moral authority but is rooted in contingent cultural processes characterized by the pursuit of particular moral and cultural values. These moral and cultural values determine the nature and limits of standard bioethics and supply the reasons for its plausibility among post-traditional people in the west. They also require reconsideration of the compatibility of American bioethics and Japanese culture.

Kazumasa Hoshino (ed.), Japanese and Western Bioethics, 73–87.

By standard bioethics I mean the family of secular approaches rooted in the theories and principles of analytical moral philosophy that are dominant in the English-speaking world. I begin by showing how the two usual positions regarding the roots of bioethics do not account for the nature and content of standard bioethics. I then adapt Charles Taylor's reconstruction of the origins of the modern identity in order to provide a superior alternative account of the roots of bioethics – one that requires a different approach to comparative bioethics in general and the comparison of American and Japanese bioethics in particular.

I. BIOETHICS AND THE RISE OF TECHNOLOGY

Bioethicists trained in moral philosophy often assume that their field of study is the product of the remarkable technological developments of recent decades that have irrevocably changed the way medicine is practiced. In its simplest form, this argument appeals to the moral quandaries of modern technological medicine which, because they are believed to be unprecedented, render traditional medical and religious ethical systems obsolete. Technology, so the argument goes, has made it possible to intervene into natural processes in ways these religious and medical traditions never anticipated, producing moral dilemmas which they are incapable of resolving. As a result we are forced to make moral choices about matters on which they can provide no guidance. A new, philosophically grounded bioethics is therefore necessary. The argument is that now that there is such a bioethics, religious and medical traditions can be dismissed as relics of the past, to be replaced by the application of common moral principles or casuistical techniques to these unprecedented problems.

This is a familiar way for bioethicists, physicians, lawyers, and policy experts to account for the rise of bioethics in the United States (Emanuel, 1991, pp. 9–14) and in Japan (Nakagawa, 1988, p. 56). If successful, this argument would not only establish standard bioethics as rationally necessary, it would also lead us to expect that Japanese bioethics will follow the American model. But I believe the argument is mistaken. Why it is mistaken can be illustrated with reference to an event frequently mentioned in support of this view: the 1968 declaration on brain death by the Ad Hoc Committee of the Harvard Medical School to Examine the Definition of Death. The declaration was a response to moral difficulties occasioned by the capacity to sustain respiratory functions by mechanical ventilation and thus seems to supply a paradigm instance of technology creating an unprecedented problem requiring a novel solution. But the declaration and its place in the ongoing debate on brain death also provide evidence against this assumption. First, however novel are the problems posed by new technology, they are not always entirely unprecedented. Concern about how to define death has a long history in medicine (Lock and Honde, 1990) and in many religious traditions. The same is true of many other "unprecedented" issues. Second, religious and medical traditions are capable of extending traditional insights to the new problems posed by technological medicine just as in previous eras they extended traditional insights to political and economic changes. Some of these traditions, including rabbinic Judaism, have arrived at viable positions on brain death without using the principles or methods of secular bioethics (Bleich, 1989; Jakobovits, 1989). Third, as indicated by the continuing controversies over brain death and by the differences in the way brain death has been received in the United States and Japan, it is often impossible to generate moral principles and apply them to cases without involving oneself in deeply held cultural or religious beliefs and practices (Zaner, 1988; Lock and Honde, 1990). If the issue of brain death is representative, as I believe it is, then the moral dilemmas generated by new medical technology do not require standard bioethics as a rationally necessary response.

II. BIOETHICS AND THE CRISIS OF MORAL AUTHORITY

If technology itself does not require standard bioethics, one may ask how within such a short period of time standard bioethicists successfully claimed authority over the resolution of the issues raised by technological medicine both in the clinic and in public policy. This brings us to a second account of the rise of bioethics, which refers not to the advance of technology but to a crisis of moral authority. As an explanation of the rise of bioethics, it has two advantages. First, it addresses the loss of the moral authority of the medical profession in the United States, which was necessary for bioethics to gain moral authority, and of religious traditions, which accounts for the nature of the bioethics that gained this authority (namely, its secular and allegedly nonsectarian character). Second, it properly identifies the grounds of the claim to moral authority in bioethics, namely its confidence that it represents and articulates a common morality.

Although the history of the American bioethics movement is in its infancy there nevertheless seems to be some consensus that the movement originated when the medical profession began to turn over some of its moral authority to outsiders. This constituted a definitive break with a long tradition according to which medical ethics was the exclusive domain of the doctor (Rothman, 1990, pp. 185–187). There are various accounts of what led to this abdication of authority. Albert Jonsen argues that it began in the early 1960s in Seattle when the limited availability of a new procedure, chronic hemodialysis, "led to a radically new solution:" the delegation of the task of selecting patients to a committee composed primarily of non-physicians (Jonsen, 1993). Jonsen, a philosopher, is interested in showing how the Seattle dialysis case marked the beginning of the "new medicine" in which issues such as microallocation, which involve populations rather than individual patients, will be central to clinical practice (Jonsen, 1990). Jonsen apparently believes that most of the ethical issues raised by new technology were in principle resolvable within the contours of an ethic internal to medicine, despite the de facto involvement of outsiders. The reason is that they were primarily questions about the proper treatment of individual patients. Not so the problems raised by the "new medicine," which includes genetics, disease prevention, measures of quality and futility of care, etc., all of which are population based rather than patient based. The ethic internal to the medical profession has no answers for this new kind of problem.

In fact the internal ethic of medicine did not successfully resolve its problems on its own; as Jonsen himself recognizes, third parties – in the form of bioethicists, patient advocates and, ultimately and most decisively, the courts – were involved early and often. From this perspective the crisis of authority derives not from the intrinsic limits of the internal ethic of medicine but from a breakdown of the confidence of outsiders in that ethic itself and/or those who represent it. David Rothman argues that the abdication of authority began when the crisis over the ethics of human experimentation precipitated by Dr. Henry Beecher's exposure of unethical research practices brought philosophers, legislators and others into the arena of clinical research (Rothman, 1990). Rothman, a social historian, links the crisis in human experimentation to the larger rights movement of the time that curbed the discretion of constituted authorities to act in the supposed best interests of others. The linkage was assured "largely because the great majority of research subjects were minorities, drawn from the ranks of the poor, the mentally disabled, and the incarcerated" (Rothman, 1991, p. 10).

Despite their disagreements, therefore, Jonsen and Rothman both point to the emergence of bioethics in a crisis of moral authority that challenged either the competence or the right of the medical profession to decide all ethical issues in the practice of medicine, and that replaced an internal medical ethic with a very different kind of ethic. What kind of an ethic was this? When persons outside the medical profession first began to claim some authority to judge ethical issues in medical practice, they proceeded in a familiar American way: by invoking a common morality allegedly shared by everyone against a parochial morality accessible only to a privileged few. Of course, "common morality" is an ambiguous term. It can refer to a set of universal moral beliefs or principles grounded in reason or intuition, or to a philosophical reconstruction of the unsystematic, largely customary moral beliefs shared by members of a community or society. In recent years many bioethicists have shifted from the first to the second sort of account (cf. Beauchamp and Childress, 1994). In either case, the argument was (and is) that bioethics represents and articulates a common morality capable of managing the new capacities of medicine.

Standard bioethics appeals to a common morality not only in contrast to a parochial professional tradition but also in the hope of overcoming the diversity and disagreement that is associated with religious traditions. In this context, the commonality is sought in secularity (or at least a limited consensus or convergence of various religious and secular beliefs), a nonsectarian posture, and standards of rationality or reasonableness that allegedly transcend or are shared by particular communities.

While few persons who accept the fact that they live in a pluralistic society would want to yield public moral authority to a particular religious tradition, the embarrassing fact remains that bioethicists do not agree on either the method or the substance of their allegedly common morality. The result, as H. Tristram Engelhardt argues, is that standard bioethics is infected with the same irreducible diversity and endless disagreement as the religious bioethics it seeks to distinguish itself from. Moreover, this is a necessary result for any secular, nonsectarian ethics that proposes content, because moral content is impossible without particularity (Engelhardt, 1995). The tenacity with which standard bioethicists cling to their claim to articulate a common morality in spite of their disagreements testifies eloquently to the modern anxiety about moral unity in the face of diversity and to the hope that a secular rationality could supply that unity where religion failed. However, the failure to arrive at a common yet substantive morality is striking evidence against the assumption that standard bioethics is a rationally necessary response to the crisis of moral authority.

But even if standard bioethics had succeeded in articulating a common morality in response to the crisis of moral authority in the West, Japanese bioethics faces a very different problem of authority. Although some Japanese traditions (notably the Jodo Shin and Nichiren traditions and their modern offshoots) claim exclusive authority, and although at times "foreign" religions were either banned (such as Christianity during the Tokugawa era) or regarded with contempt (as was Buddhism by some nationalistic groups during the Meiji period), in general both Japanese culture and individual Japanese persons have traditionally blended various religious traditions into a common cultural whole or a unified life history. This means that religion in Japan is not as marked by rival traditions with their explicit disagreements or as sharply distinguished from the broader culture as it is (at least in principle, though less often in practice) in the contemporary United States. The need for a common morality, therefore, is less likely to be expressed in terms of sharp distinctions between a common morality and the moralities of medical or religious traditions. Of course, Japanese society is changing rapidly, but it is a mistake to assume that the changes will inevitably reproduce the same moral and institutional patterns that analogous changes have produced in the west. If Japanese bioethics follows the precedent of other traditions Japanese have adapted to their own dynamic yet unified cultural patterns, one should instead assume that values such as self-determination and beneficence will blend in with perpetual (though changing) Japanese concerns with the family system, purification, and harmony with society and nature that are among the chief themes of Japanese religion(s). I believe that a careful reading of the Japanese contributions to this volume supports this assumption rather than the assumption by some of the American contributors that there are no significant differences between common morality in the two cultures.

III. BIOETHICS AND THE LOSS OF TRADITION

Turning back to the West, the second account gets the issue of moral authority right but ignores the deeper roots of the crisis of moral authority. As a result, it fails to account for the basic moral content shared by standard bioethicists in spite of their disagreements, for the exact grounds on which traditional forms of ethics are considered inadequate, and for the kinds of issues standard bioethics is unable to address because of its moral commitments. The deeper roots of the crisis of moral authority involve the loss of tradition in the west. The loss of tradition means the loss of a certain moral vocabulary – one that places the pursuit of health in the context of the pursuit of a good life within the limits set by fate or necessity – and its replacement by a new moral vocabulary – one that is dedicated to eliminating suffering and expanding the range of human choice. This locates the roots of bioethics in the emergence of modern moral theories.

In order to describe this loss of tradition I will sketch some rather formal features of medical and religious moral traditions. The accounts are formal – I do not claim that they describe any historically actual medical profession or religious community. Nor do I claim that premodern societies were marked by practices that conformed to a unified moral vision that is now lost. Instead I offer these accounts as instances of a kind of moral vocabulary that characterizes traditional ways of life but that modern moral discourse disavows.[1]

An adequate description of medicine as a traditional practice would clarify the relation between knowledge and technical skill, on the one hand, and health as an end for the particular patient being treated, on the other hand. As such, medicine is a practical art. It assumes an understanding of health as a standard of bodily excellence or "an activity of the living body in accordance with its excellences" (Kass, 1985, p. 174). But it also recognizes that this standard must be specified with reference to each person, so that the task is to determine the nature and degree of health appropriate for a particular patient. Hence medicine requires general knowledge about excellent bodily functioning, insight into the relation of this functioning to the capacities and roles of a particular patient, and awareness of the possibilities and limitations of facilitating or restoring functioning for this particular patient. It is impossible to achieve competence without this threefold knowledge because one can identify a skilled practitioner only by her ability to fulfil the possibilities and observe the limitations of bodily health for a particular patient.[2]

Three points are especially important here. First, the standards and ends of medicine are internal to the practice itself: medicine is not a set of technical skills in the service of ends that can be described apart from standards of excellence of bodily functioning, though what this excellence is for any given patient involves more than the body. Second, these standards need not and will not be static. Hence there is room in principle for technological change and for growth of insight. As with any practice, "conceptions of goods and ends which the technical skills serve . . . are transformed and enriched by these extensions of human powers and by that regard for its own internal goods" (MacIntyre, 1984, p. 193). But, third, this transformation and enrichment will occur within the recognition of health as a mortal good and of human beings as destined to suffer disease and die (Kass, 1985, p. 163).

I now turn to the characteristics of the moral authority of a religious tradition. These characteristics may or may not be explicit; for some persons they may be almost entirely customary. An adequate description of such a tradition would include its account of the nature and proper ends of human beings, and of the virtues that either constitute those ends or enable one to attain them. It would also include an account of how one attains the proper ends of life, the obstacles (both internal and external) one encounters, the authorities (official or unofficial) that lead and instruct one, and the powers (for example, divine grace) that assist one. Such a tradition will also include norms, rules and prohibitions. These specify actions or modes of conduct that are either required or ruled out in order to engage in this way of life. For example, both the major precepts of the natural law in Thomas Aquinas and the five moral precepts in Theravada Buddhism specify the necessary conditions for embarking on a way of life devoted to reaching the higher ends in each tradition. One justifies these precepts by showing how that way of life depends upon observing them. Without some such norms and prohibitions it would be impossible for that way of life to be the distinctive form of belief and practice that it is. Moreover, from within such a way of life, these norms, rules and prohibitions will have a casuistical framework in which they are refined, contested and sometimes abandoned; conflicts between them are resolved; and rules and authorities for interpreting them are identified and contested. I elaborate these features of norms, etc., because they are treated differently when severed from their location in a tradition.

Such a tradition will possess two characteristics relevant to health and illness. First, it will provide an account of how bodily health is related to the ends of life and what degree of health is necessary to attain those ends. Second, when technology brings new areas of bodily life into the realm of medical intervention, both those ends themselves and certain norms and prohibitions will place limits on the pursuit of health and the means by which it is pursued.

These characteristics of medical and religious traditions are absent in the modern moral framework that challenged these traditions. This modern framework, or so I argue, accounts for the cultural content on which bioethics draws. Charles Taylor's reconstruction of the sources of the modern moral self offers one fruitful way of understanding this framework (Taylor, 1989). What follows relies heavily on key elements of Taylor's interpretation, though I alter it at some points, add to it, and apply the resulting product to medicine. One of the chief characteristics of the modern moral framework, according to Taylor, is the moral valuation of ordinary life. For Protestant Christianity human effort is fruitless in attaining salvation, which comes through divine grace alone. Hence rather than directing one's life toward the attainment of moral and religious perfection, human effort is to be directed toward serving the needs of one's neighbors. This is done by engaging in the pursuits of ordinary life such as family and work. But if the needs of one's neighbors are to be met, one's work must be disciplined and effective. It became clear beginning with Francis Bacon that effectiveness would require an instrumental approach to nature, ultimately including human nature, in order to fulfil its moral project. In this spirit Bacon praised the mechanical arts and disparaged speculative science for doing nothing "to relieve and benefit the condition of man" (Bacon, 1960, pp. 71–72; cf. Taylor, 1989, p. 213).

The instrumental approach to nature was supported by a theological conviction that God has ordered nature for the preservation and enhancement of human life. Nature is therefore governed by divine providence as in the Stoic and medieval cosmos, but the conception of a providential order has changed. The ancient and medieval conception of nature as a teleological order from which a hierarchy of ends could be derived was replaced by the burgeoning conception of nature as a law-governed mechanism, susceptible to human control and neutral with regard to ends – an order, therefore, which permits human control for the purposes of human preservation and well being.

Up to this point, the roots of modern morality are expressed within Protestant Christianity. But as Taylor emphasizes, radical Enlightenment thinkers such as Jeremy Bentham were able to understand their secular agenda as a superior way of affirming ordinary life and expressing benevolence. According to them, the affirmation of ordinary life meant being true to the demands of ordinary human nature and so identifying good with pleasure and evil with pain. The Protestant commitment to meeting the needs of the neighbor now became a set of obligations to prevent and remove the causes of pain and to maximize the quantity of pleasure. As Taylor argues, this made it possible for the first time to put the relief of suffering (and the avoidance of cruelty) at the center of the social agenda. This emphasis on the relief of suffering in turn resulted in a new standard for all remaining conceptions of religious, moral and legal order: do they lessen the amount of suffering in the world or contribute to it (Taylor, 1989, p. 331)? From now on all these conceptions of order would have to present their credentials for relieving suffering to gain admission to the moral realm, credentials few such conceptions could produce.

Not surprisingly, this new moral agenda was closely connected with the loss of the belief in divine providence that had sustained the Protestant moral enterprise. Ever since the nominalists it had been difficult to support belief in divine providence on philosophical grounds, and as the mechanistic explanation of nature reached its climax with Newton, providential and the remaining teleological approaches to nature were both discredited. Confidence in a providential order therefore gave way to a growing emphasis on the need to extract the preservation and enhancement of human life from an indifferent nature by means of technological labor. A point Taylor does not emphasize is that the loss of ideas of providence or a meaningful cosmic order removes the incentive to find any religious or cosmic meaning for suffering. Moreover, the mechanization of nature means that suffering from natural causes is no longer an inevitable feature of the world but is, to the extent that human beings are capable of controlling nature, an object of human responsibility. Hence the new world view both requires the elimination of suffering and makes it possible.

The contrasts with traditional ways of life are clear. First, the meaning of bodily life, which was once determined by an account of its excellent functioning and limited by its subjection to fortune, will now be determined by its susceptibility to technological control. The medical wisdom of learning the limits of healing and accepting the mortality of the body will yield to Bacon's admonition to call no disease incurable and, even more presciently, to orient medical knowledge to the prolongation of life (Bacon, 1894, pp. 163, 166–168). Second, the concern with the preservation and enhancement of ordinary human nature combined with the concern to relieve suffering means that health will become an end in itself rather than a condition or a component of a virtuous life. Medical care will be devoted to relieving and eliminating suffering wherever it is found rather than to the management of health for the pursuit of virtue. Third, rules and prohibitions limiting what can be done to the body to relieve suffering will appear to be at best insufficiently concerned about suffering and at worst arbitrary and even cruelly insensitive.

This combination of technological control over nature (including the human body) and a moral commitment to relieve suffering by preventing the harms and eliminating all the conditions and limitations that threaten bodily life accounts for a large part of the nature and task of medicine in the modern era. But there is one more chapter to the story. A second aspect of the modern moral framework is what Taylor calls inwardness. Inwardness has deep Augustinian and Cartesian roots, but during the Romantic period it surfaced in the inner conviction of the importance of one's own natural fulfillment. The idea is not only that each individual is unique and original but that this uniqueness and originality determines how he or she ought to live. There is an obligation (more aesthetic than moral) for each person to live up to his or her originality (Taylor, 1989, pp. 370–376). What follows from this is the importance contemporary moderns place on free self-determination. Together with the ideal of universal benevolence, self-determination also leads to the idea of the subject as bearer of rights of immunity and entitlement. From this follows expectations that the expansion of the reign of technology over the body should be accompanied by, and in fact should make possible, the expansion of the reign of human choice over the body, and that medicine should enable and enhance whatever pattern of life one chooses.

Taylor argues that the Victorian era brought together these Enlightenment and Romanticist trends and bequeathed them to us – along with a view of history as a story of moral progress over our forebears, a progress marked by our greater sensitivity to and eradication of suffering and our greater latitude for human choice. This view enabled the Victorians to be convinced of their moral progress over the age of religion even as it enables their successors in this century to be convinced of their moral superiority over the Victorians (Taylor, 1989, pp. 393–396). The result is that medicine is based on practices and techniques of control over the body rather than wisdom about the body. The task of public policy is to negotiate rights of immunity and entitlement rather than to determine the place of health, illness and medical care in a well-lived and responsible life and in a good community. Traditional moral injunctions that limit or inhibit what medicine can do appear arbitrary, but there is no broader framework to evaluate and criticize the commitments of modern medicine. In the absence of such a framework, the commitment to relieve all suffering combined with an imperative to realize one's uniqueness leads to cultural expectations that medicine should eliminate whatever anyone might consider to be a burden of finitude or to provide whatever anyone might require for one's natural fulfillment. This does not mean that individual conceptions of this burden or this fulfillment are necessarily arbitrary. But it does mean that modern moral frameworks provide no vocabulary with which to deliberate about what makes some such conceptions better or worse than others.

IV. CONCLUSION

This brief sketch of the modern moral framework allows us to identify the major cultural values that standard bioethics draws upon. But two points must be made clear. First, I do not believe that there once was a golden age when medical care was grounded in a robust view of the good or that individual choices now are purely arbitrary. Nor do I believe that it is possible or desirable to reverse the technological revolution in medicine and simply return to traditional ways of life. Still less do I believe that publicly enforced consensus about these matters is possible or desirable. My argument is the more modest one that modern moral frameworks provide no vocabulary with which to deliberate about the meaning of corporeality, what bodily excellence is, what goods health should serve, or what limits the control of our bodies by technology should observe. Second, my account simply identifies some features of the modern moral framework and does not do justice to the rigor with which bioethicists have articulated and balanced these features.

Nevertheless, this account supports three conclusions. First, standard bioethics is not a rationally necessary response to technology or to the need for a common morality, but is rooted in certain cultural processes that lend plausibility to the characteristics that define standard bioethics. Those who adhere to traditional ways of life will find neither the methods nor the conclusions of standard bioethics rationally compelling. Second, standard bioethics is not neutral toward particular views of the good, but articulates its own thin view of the good whose features (the commitments to the elimination of suffering and the expansion of choice) I have articulated. Far from being neutral, this thin view of the good is at worst hostile to robust views of the good and at best unable to resolve pressing questions regarding the nature and limits of technological control of the body. Third, this account provides the grounds for a more adequate comparison between American and Japanese bioethics. My argument has been that American bioethics does not rest upon culture-transcending rational grounds but on certain cultural processes stemming from the loss of tradition – processes that replaced views about the finitude of the body and its place in a morally worthy life with technological control of the body for the purposes of eliminating suffering and expanding choice. This raises two questions for Japanese bioethics. First, are there analogous values in Japanese culture that also support efforts to eliminate suffering and expand the realm of choice over that of fate? And second, are there values in Japanese culture that place these efforts in a broader moral framework that makes possible substantive moral judgments about the nature and limits of such efforts?

To answer questions like these, comparative bioethics must answer questions about the meaning of bodily life, the relation between humanity and nature, the meaning of suffering and the relation of health and illness to the good life. Some of the Japanese essays in this volume contribute to such a project. But standard bioethicists will be able to join them only when they divest themselves of their false assumptions about the culture-transcending rationality of their enterprise.

Department of Religious Studies
Rice University
Houston, Texas U.S.A

NOTES

[1] Both accounts reflect my own understandings of what medical and religious traditions are. However, my discussion of medicine as a practice is informed by descriptions of medicine by Leon Kass (1985) and by Edmund Pellegrino and David Thomasma (1981), and my discussion of the place of health and illness in religious traditions is informed by the views on the relation of health to virtue in Moses Maimonides (1975) and by John Bowlin's account of the relation of health, fortune, and virtue in St. Thomas Aquinas (Bowlin).

[2] I leave aside the crucial question of whether medicine as a tradition can be complete without an adequate conception of what capacities and roles are conducive to a good or morally worthy life. However, I am convinced that medicine as a tradition is dependent on such a conception, and is therefore dependent on religious or moral traditions. In a response to an earlier draft of this paper, Robert Veatch remarked to me that the notion of medicine as an independent tradition is a historical fiction. If I am right, then it is also a philosophical mistake.

BIBLIOGRAPHY

Bacon, F.: 1894, *Works of Lord Bacon*, J. Devey (ed.), George Bell and Sons.

Bacon, F.: 1960, *The New Organon and Related Writings*, F.H. Anderson (ed.), Liberal Arts Press, New York.

Beauchamp, T. and Childress, J.: 1994, *Principles of Biomedical Ethics*, 4th edition, Oxford University Press, Oxford.

Bleich, J.D.: 1989, 'Of cerebral, respiratory and cardiac death', *Tradition* **24**(3), 44–66.

Bowlin, J.: 'Health, fortune and moral authority in medicine', unpublished paper.

Emanuel, E.: 1991, *The Ends of Human Life*, Harvard University Press, Boston.

Engelhardt, H.T., Jr., 1995, 'Content, tradition, and grace: Rethinking the possibility of a Christian bioethics', *Christian Bioethics* **1**(1), 29–47.

Jacobovits, Y. (trans.): 1989, '[Brain death and] heart transplants: the [Israeli] Chief Rabbinate's directives', *Tradition* **24**(4), 1–14.

Jonsen, A.R.: 1990, *The New Medicine and the Old Ethics*, Harvard University Press, Boston.

Jonsen, A.R. (ed.): 1993, 'The birth of bioethics', *Hastings Center Report*, Special Supplement **23**(6), 1–4.

Kass, L.R.: 1985, *Toward a More Natural Science*, Free Press.

Lock M. and Honde C.: 1990, 'Reaching consensus about death: Heart transplants and cultural identity in Japan', in G. Weisz (ed.), *Social Science Perspectives on Medical Ethics*, Kluwer Academic Press, Dordrecht, pp. 99–119.

MacIntyre, A.: 1984, *After Virtue*, 2nd edition, University of Notre Dame Press, Notre Dame.

Maimonides, M.: 1975, *Ethical Writings of Maimonides*, R.L. Weiss (ed.) with C. Butterworth, New York University Press, New York.

Nakagawa, Y.: 1988, 'Development of bioethics', in J. Bernard, K. Kajikawa and N. Fujiki (eds.), *Human Dignity and Medicine*, Excerpta Medica, pp. 55–61.

Pellegrino E. and Thomasma, D.: 1981, *A Philosophical Basis of Medical Practice*, Oxford University Press, New York.

Rothman, D.J.: 1990, 'Human experimentation and the origins of bioethics in the United States', in G. Weisz (ed.), *Social Science Perspectives on Medical Ethics*, Kluwer Academic Press, Dordrecht, pp. 185–200.

Rothman, D.J.: 1991, *Strangers at the Bedside: A History of How Law and Bioethics Transformed Medical Decision Making*, Basic Books, New York.

Taylor, C.: 1989, *Sources of the Self: The Making of the Modern Identity*, Harvard University Press, Boston.

Zaner, R.: 1988, *Death: Beyond Whole-Brain Criteria*, Kluwer Academic Press, Dordrecht.

KEVIN WM. WILDES, S.J.

SANCTITY OF LIFE: A STUDY IN AMBIGUITY AND CONFUSION

Bioethics is a field that has defined itself in moral controversies. For example, bioethics has emerged as a discipline in the attempts to resolve moral controversies surrounding medical practices and health care policies in areas such as experimentation and research, abortion, reproduction, and the allocation of resources in health care. One source of constant moral controversy has been the issues surrounding death and dying. There have been controversies about the definition of death, the extent of the obligation to treat the dying, the use of resources for the care of the dying, euthanasia, and assisted suicide. From the celebrated cases in the United States of Karen Ann Quinlan and Nancy Beth Cruzan to the ruling on assisted suicide in the state of Washington (see, Compassion in Dying) the controversies of death and dying have been part of bioethics and public policy.

In the debates over moral issues in death and dying terms like "sanctity of life," "respect for life," and "human dignity" have been deployed often. These terms have been appealed to as men and women have sought to create public policy and develop moral consensus in the controversies of death and dying. Such terms are often used as the final court of appeal to justify particular moral choices or public policy in health care. Yet such terms are often heterogeneous in their meaning. Each of these terms is filled with enough ambiguity so as to bring together a wide range of hopes, images, feelings, and values that make the meaning and use of the terms very different. As a consequence people can appeal to the same term and draw very different conclusions about what should be done. In the case of Nancy Beth Cruzan one finds different parties in the controversy holding very different opinions about the case and yet each appeals to the concept of "sanctity of life."

In this essay I want to examine, in a limited way, some of the different meanings of these terms. The first section of this paper will examine

Kazumasa Hoshino (ed.), Japanese and Western Bioethics, 89–101.

how terms such as sanctity of life are used within different religious traditions. In the second section I will outline how some have tried to use these terms in general, secular bioethics. It will be argued that as these terms are progressively excised from their religious or other specific context they become so ambiguous as to be meaningless. I will illustrate this point by examining the use of such terms in the case of Nancy Beth Cruzan. The principal argument of this presentation will be that terms like sanctity of life and respect for life can only be understood within the context of a moral framework and language. When excised from such frameworks the terms become incapable to direct health care policy in secular, morally pluralistic societies and they ought to be discarded.

I. RELIGIOUS PERSPECTIVES AND THE SANCTITY OF LIFE

One finds terms such as "sanctity of life" and "respect for life" used in different religious traditions. The assumption is often made that the terms are more or less equivalent in meaning. However, after examining how either of these terms are defined within a tradition it becomes clear that they are not interchangeable. While they may have a family resemblance one will not, necessarily, draw the same conclusions from their different uses in different traditions. In the Christian view it is God who is the source of life's sanctity and human life has a unique role in the created world because of its relationship to God. This view, focused on human life, is quite different from the Buddhist view that all life is intrinsically sacred. Indeed one finds that even within different religious traditions the terms take on different meanings. A consequence of such ambiguity in meaning, within and between moral traditions, is that terms like sanctity of life often convey very different meanings and justify different choices of action.

A Christian Perspective

While different meanings can be associated with "sanctity" the meaning that seems central to the term sanctity of life is the meaning of "being hallowed or sacred." This definition conveys the notion of inviolability which is what the latin root of sanctity ("sanctitas") means. In the Christian tradition the claims about the sanctity of life seem to communicate the supposition that *human* life has an inviolability or a sacredness. One function of the claim that human life is sacred is to direct our actions. This claim of sanctity put limits on what can be done (e.g., human life ought not be taken) and makes demands on what should be done (e.g., life ought to be preserved).

It is true that a fundamental and traditional concern of Christianity has been a concern not to harm human life. For example, *The Didache*, which dates to the first century, enjoins Christians: "[T]hou shalt not procure abortion nor commit infanticide" (Didache I). It also condemns the murder of children (Didache V).[1] While traditional Christianity has held specific prohibitions against killing, it is too much to claim that these prohibitions form a doctrine of the sanctity of life. Joseph Boyle has argued that the phrase "sanctity of life" represents a family of values that is rarely articulated carefully even in the Christian tradition (Boyle, 1989). Two themes seem to be interwoven in the Christian uses of "sanctity" when talking about human life. The first theme is that such life is holy because life is a gift from the Holy. In Christianity the origin of human sanctity is God. Life is a gift from God, who is Life, to human beings. However, this alone does not explain the Christian tradition insofar as *all life* is a gift from God and yet human life has a unique place in the created world. A second theme, or necessary condition, is that human life must have a special relationship with the Divine that sets human life apart from other forms of created life.

The uniqueness of human life, for Latin Christianity, has been explored in reflections on man as the "imago Dei" (Genesis 1:26). That is, human beings are made in the image of God. The Fathers of the Church were influenced by the view of the Old Testament that emphasized God as both the beginning and the destiny of man. Since *all* life is created, what is it that distinguishes human life as "sacred?" Human life has a "unique" status in that God impresses onto the human person God's own image and resemblance and therefore makes the human being above other beings which are God's creatures but not mirrors of the Creator. Human beings are part of the creation but they are distinguished from the rest of the creation as they are to rule as God rules. All life, since it comes from God, has a sacredness about it and demands respect, for it belongs to an-Other. The special dignity and sanctity of human life comes from bearing the image of God and the responsibility to rule like God.

The divine is expressed in the world in the human. Irenaeus best captured this patristic sense when he wrote: "Gloriam enim Dei vivens homo, vita autem hominis visio Dei" (Irenaeus, 4. 20. 7). The glory of God is the living human and the life of the human is the vision of God. The "glory of the human" is not a modern sense of self-improvement but the expression of the Divine in the human. Human life is made "holy" and endowed with "sanctity" and "dignity" because of its relationship to God.

Throughout the history of Christian theology and spirituality the meaning of human sanctity and dignity has been developed in different ways. For example, in the reflections of twelfth and thirteenth century Latin theologians there is a search to identify the characteristics that distinguished human life from all other forms of bodily, created life. The epoch of scholastic philosophy was characterized by an emphasis on the *intellectual and rational* dimension of God's image impressed onto the human. The expression "imago in specula rationis," commonly used by scholastic thinkers, manifests this view. The pivotal point in this interpretation is that human sanctity consists in the human ability to know himself and God. The faculties of intellect and will, unique to man in the embodied, created world, were seen as the most divine of human attributes.

In contrast, theologians in the twentieth century have sought a less rationalistic and more balanced, integrated view of the human person. Many theologians have sought to develop a Christian anthropology that situates the rational within the relational and social context of human life (Rahner; Schotsmans; Wojtyla). In contemporary Latin theology one finds that these two conditions for sanctity of human life have often been blended together under a theme of "stewardship." As life is a gift and since it is a gift with a purpose, one is called to be a good steward of the life given. The two conditions for the sanctity of human life lead to different moral imperatives. First there are proscriptions against the taking of innocent human life. For example, according to the Roman Catholic tradition innocent human life can not be directly, intentionally taken. Thus there are proscriptions against suicide, abortion, and murder. At the same time the tradition has reflections not only about the protection of life but that it should be used well. Archbishop Daniel Cronin writes: "Among the natural gifts with which the Most High God have favored man, there is none so excellent as that of life, because it is life that is the basis for all else that man has or can hope to attain" (Cronin, p. 3). In this vein there are serious and prolonged reflections on the positive obligation to preserve life insofar as it is a basis for achieving other goods (see Cronin, 1958).

As one examines particular moral questions surrounding death and dying one discovers that the term "sanctity of life" has contributed very little to their discussion in the Christian tradition. Issues about suicide and euthanasia have been treated by the prohibition against the direct, intentional taking of innocent human life. The questions about the obligation to preserve life by pursuit of life sustaining treatments have been addressed by the distinction of ordinary and extraordinary means. Questions about the definition of death have been addressed by discussion of the metaphysical questions of hyelemorphic composition and whole brain death. In no one of these areas has the term sanctity of life contributed to the substantive discussion. Rather, the term seems to have emerged more in secular discussions of bioethics and in the abortion debate (see, Brody).

Buddhist Perspectives

The ambiguities in terms like sanctity of life are made ever more clear when one contrasts a Christian tradition with other religious traditions. In Buddhism the first of the Grave Precepts is to "affirm life; do not kill." The principle of respect for life, as it has been called, is one of the foundations of Buddhist ethics (Fujii). It has been the basis for a clear-cut position against practices such as abortion (Nolan, p. 194; Stevens 138–139). However, in Buddhist thought the principle of respect for life must be understood within the context of other aspects of Buddhist teaching as well as the other precepts. Different traditions within Buddhism balance the concern for respect for life with concern with doing "the most compassionate action" (Nolan, p. 194). So while there is a general prohibition against a practice such as abortion, for example, such particular decisions must be made within the context of the other elements of suffering and with a view toward compassion. A decision to violate the first precept is one that should not be taken lightly but to fail to violate it, when compassion demands, it is to generate negative karma (Kaplean, p. 228; Rinpoche, p. 376). One finds then that the precept of respect for life needs to be interpreted within the framework of the situation and other moral demands. One also finds that the different traditions of Buddhism will make different interpretations of how to deploy the Precept.

In addressing the questions of treatment decisions at the end of life one finds different interpretations of the first precept. For some Buddhists following the first precept means to utilize whatever means of treatment and recovery are available. The argument is that human life offers an opportunity to transcend suffering through enlightenment and there is the possibility for every disease to be cured as long as life continues (Ratanakul, p. 310). However, the Buddhist discussion of the issue needs to balance the first precept to respect life with the demands of suffering and compassion. Treatment cannot be refused simply to escape suffering but one can refuse treatment for selfless and compassionate reasons. That is, a person may refuse treatment to act compassionately towards family and friends, or to relieve burdens (e.g., economic) on them.

One finds similar sets of issues and themes arising in discussions of active euthanasia.[2] Those who disapprove of forgoing treatment in that it has the character of suicide or "death-seeking" oppose active euthanasia (Nolan, p. 199). Yet, as one might expect, those who allow an exception do so because they appeal to compassion. What is crucial for my argument is that the precept of "respect for life" is balanced with compassion and it is this balancing that helps determine the significance of the precept. This balancing only takes place within the context of a tradition.

Summary

This brief overview illustrates how terms like sanctity of life or respect for life have been understood within different moral narratives. In the Christian tradition *human* life is sacred because it bears the divine image, while in Buddhism human life is sacred because all life is sacred. However, what is perhaps most instructive for general, secular bioethics is that even within these narratives there are ambiguities in understanding these terms in light of particular moral choices. As one examines the use of such terms, these ambiguities offer a warning for secular bioethics. These terms, which are difficult to define within a moral tradition, will take on so many meanings in general, secular discourse as to become meaningless. The *Cruzan* case illustrates how people with very different views of moral controversies in death and dying can reach very different conclusions by appealing to the "sanctity of life."

II. BIOETHICS AND SANCTITY OF LIFE

General, Secular Bioethics[3]

Discussions of sanctity of life seem to have entered the English bioethics literature in the early 1950's.[4] In 1957 one finds Glanville Williams using the term "sanctity of life" in some of his jurisprudential writings (Williams). The use of the term initially centered around issues such as euthanasia and abortion. In criticizing the legalization of voluntary euthanasia John Bonnell argued that Christianity has emphasized "the sanctity of human life and the value of the individual, even the humblest and lowliest, including the afflicted in mind and body" (Bonnell, 1951). The article by Bonnell was in part a response to an essay by Joseph Fletcher in which Fletcher argued for the centrality of persons over mere life. Fletcher argued that one would be better served to speak of the role that persons have in deciding for themselves rather than to appeal to principles like "sanctity of life" (Fletcher, 1951).

In 1964 Norman St. John-Stevas argued that the Christian attitude toward euthanasia is based on "the principle of the sanctity of life" (St. John-Stevas, p. 43). In these early uses of "sanctity of life" in bioethics there are clear religious (particularly Christian) presumptions (Ramsey, 1967). Harmon Smith, in a commentary on Paul Ramsey, speaks of the religious framework within which the notion of sanctity of life was understood (Smith, 1970). He writes:

> ... the question of *when* sanctity attaches to human life is not religiously problematic at all (for Ramsey): "One grasps the religious outlook upon the sanctity of human life only if one sees that this life is asserted to be *surrounded* by sanctity that need not be in a man; that the most dignity a man ever possesses is a dignity alien to him.... A man's dignity arises from God's dealings with him, and not primarily in anticipation of anything he will ever have it in him to be" (Smith, p. 42).

Sanctity of human life for Ramsey and other Christians is derived from the fact that God values human beings.

There have been some attempts in bioethics to give the principle of the sanctity of life a less religious significance. One example is the work of Daniel Callahan in his exploration of abortion. Callahan understands the problem with using a religious term like sanctity of life in a secular society. In his attempt to give sanctity of life a secular meaning Callahan writes: "An affirmation of the sanctity of life which required that one accept a religious view of man's origin would provide a weak base upon which to build a consensus. One then would seem to be saying that there is nothing whatever upon which to ground the sanctity save that of religious belief. . . . " (Callahan, 1970, p. 315). Callahan attempts to use the content of religious, particularly Christian, views without their theological foundations. This analysis and transformation leads Callahan to understand the sanctity of life as an affirmation of a morality that affirms "the protection and preservation of human life, both actual and potential" (p. 343). Callahan articulates a diverse collection of rules gathered under the rubric of the term sanctity of life. These include: "(a) the survival and integrity of the human species, (b) the integrity of family lineages, (c) the integrity of bodily life, (d) the integrity of personal choice and self-determination, mental and emotional individuality and (e) the integrity of personal bodily individuality" (p. 327). While he recognizes the ambiguity of the term, Callahan still attempts to derive from it some useful direction and moral sense.

K. Danner Clouser, in 1972, criticized the term sanctity of life for this very ambiguity. He argued that all the different meanings, and their implications, are mixed together. Clouser wrote: "I find the sanctity of life concept to be impossibly vague and to be a concept that is inaccurate and misleading, whose positive points can be better handled by other well-established concepts" (Clouser, p. 119). William Frankena, in 1975, sorted out the different meanings that are gathered under the term sanctity of life.

1. The sanctity of bodily human life should be distinguished from that of individuality or personality. The sanctity of human life (bodily) is relevant to the discussion of questions of shortening or preventing human life.
2. Mere life, whether that of a vegetable, animal, or human organism, has no moral sanctity as such, though it may have aesthetic and other kinds of nonmoral value, and may be a necessary condition of consciousness, rationality, or morality.
3. Life has moral sanctity, but only where it is a condition of something more, as it is in human fetuses.
4. There is something inherent – consciousness, feeling, reason – in such living beings.
5. Even if the moral sanctity of human life (bodily) is not absolute, it is considerable, at least from the moral point of view, but it is only *prima facie* or presumptive.
6. The only tenable view, then, is a derivative, qualified, and noncomprehensive ethics of respect for life (p. 58).

Frankena's work points out the ambiguity of terms such as "sanctity of life" and "respect for life." His work also points out the conceptual problems to deploy such terms in a general, secular bioethics. In the midst of all these different accounts of these terms there is no way, in a general secular context, to pick out which is the correct interpretation. General, secular bioethics runs the risk of speaking in babel when terms like "sanctity of life" and "respect for life" are used. Indeed the case of Nancy Beth Cruzan, and the general issues of death and dying, make this babel very clear.

Cruzan: An Example

Despite the ambiguities in the term it has played a crucial role in certain controversies in bioethics about decisions at the end of life. The Missouri Supreme Court, for example, in its decision in the *Cruzan* case asserted that the protection of the "sanctity of life" was a *state interest* such that the state should prevent the withdrawal of feeing and hydration (Cruzan v. Harmon).

The case involved a patient who had been in a persistent vegetative state since an automobile accident. As she was anoxic for twelve to fourteen minutes Mary Beth Cruzan suffered irremediable brain damage. Subsequent to the accident a gastrostomy feeding and hydration tube was placed with the consent of her husband. When it became clear to her parents that their daughter had no chance of recovery, they sought removal of the tube. The employees of the hospital refused to comply with the request. The Supreme Court of Missouri, in a divided opinion, denied the parents' request. The state argued that there was a state interest in the sanctity of life under the *parens patriae* doctrine of common law (Payton). For this essay it is important to note that the State Supreme Court, as well as several opinions of the U.S. Supreme Court upheld "sanctity of life" as a value which trumps other values. The Missouri Supreme Court wrote:

> The State's interest in life embraces two separate concerns: an interest in the prolongation of the life of the individual patient and an interest in the *sanctity of life itself* (Cruzan, 1988, emphasis added).

In the view of the majority, these general interests are strong enough to foreclose any decision to refuse treatment for an incompetent person unless there exists clear and convincing evidence that the person previously had made such a choice.

While asserting sanctity of life as a value that orders other values its meaning is never made clear nor is there an argument as to why this value should trump other values. The opinions of the justices in these decisions, in fact, reflect a pluralism of moral vision and language. Indeed, Justice Stevens, of the United States Supreme Court, in his dissenting opinion in the appeal of the case, pointed out that: "Life, particularly human life, is not commonly thought of as a merely physiological condition or function. Its sanctity is often thought to derive from the impossibility of any such reduction" (*Cruzan*, 1990, Stevens' dissent, Part III). Stevens' dissent represents a very different interpretation of sanctity of life from the one deployed by the State Supreme Court. For Stevens, and others, the sanctity of human life is centered on the capacities of personhood. In a sense Stevens' view is not far from the view of Medieval theologians who focused on the rational capacities as the essence of the divine image in the human. The loss of these capacities ends the obligation to sustain life.

The different interpretations of the nature of the "sanctity" lead to very different outcomes of the case. Stevens captures the difficulty of using terms like sanctity of life. He points out "the more precise constitutional significance of death is difficulty to describe; not much may be said with confidence about death *unless it is said from faith*, and this alone is reason enough to protect the freedom to conform choices about death to individual conscience" (Cruzan, 1990; Stevens's dissent, Part III, emphasis added). Indeed one sees in the Cruzan case two different appeals to sanctity of life. One appeal interprets the term to require aggressive medical treatment so that she can be kept alive. The other interpretation concludes that the treatment should be withheld.

III. CONCLUSION

The argument of this paper has been that terms like "sanctity of life" are so ambiguous that they can support starkly contrasting choices in the treatment of the dying. Outside the context of a particular moral community or moral narrative the principle of sanctity of life can be interpreted in at least the following ways.

1. The principle of the sanctity of life requires one to save human life at all costs.

2. The principle of sanctity of life requires that one preserve the values associated with human life and these can be jeopardized if one tries to save mere biological life.

In the second interpretation sanctity of life is not to be achieved simply through biological life, but through a self-conscious moral life. This is the interpretation that lies behind the traditional Roman Catholic distinction of ordinary and extraordinary means (Wildes). There is a recognition that if one makes the mere prolongation of life an overriding good, the place of other moral goods will be disturbed (Pope Pius XII).

The difficulties with appeals to terms like "sanctity of life," "respect for life," or "human dignity" is that they are more like slogans than principles outside of a particular context. I have argued that such terms can be understood within the language of a moral community. Excised from such communities such terms become ambiguous and useless. That is, they bring together a number of issues and attitudes more than set out foundations or rules for choices. People with very different views rally around them. It becomes impossible to establish a canonical interpretation of the principle of the sanctity of life.

Indeed if one looks at the Latin Christian tradition it is clear that "sanctity of life" has not been understood as a moral principle. Rather, one might see it as a background assumption that shaped moral principles such as the prohibition against directly intending to take human life or the distinction of ordinary and extraordinary means. It is only in recent years, in debating issues of death and dying, that such terms have been invoked and used as principles. However, as I have argued, when such terms are taken from their basic role they yield very little. The result is more like babel than language.

Georgetown University
Washington, D.C., U.S.A.

NOTES

[1] For an excellent overview of the Christian tradition on abortion see John Noonan, 'An almost absolute value in history,' in *The Morality of Abortion*, (Cambridge, MA: Harvard University Press, 1970), pp. 1–59.

[2] There is in Buddhism the practice of self-immolation which has not been understood as an act of euthanasia or suicide. Rather it is understood as the practice of giving one's self over and merging one's self into transitory reality (see, Fujii).

[3] This discussion of general secular bioethics and sanctity of life has grown out of

many long discussions with H.T. Engelhardt, Jr. and work that we have done together (see Engelhardt, 1996).

[5] There are publications prior to 1950 exploring "the sanctity of life" (see Hillis, 1921; Young, 1932). In bioethics there was a renewal of some of the concerns of Albert Schweitzer for a reverence for life. In the *New England Journal of Medicine* William Sperry wrote that "reverence for life" is the ethical basis of both the profession of medicine and Christian ministry (Sperry, 1948, p. 988).

BIBLIOGRAPHY

Bonnell, J.S.: 1951, 'The sanctity of human life', *Theology Today* **8**, 194–201.

Boyle, J.: 1989, 'Sanctity of life and suicide: Tensions and developments within common morality', in B.A. Brody (ed.), *Suicide and Euthanasia*, Kluwer Academic Publishers, Dordrecht, The Netherlands, pp. 221–250.

Brody, B.A.: 1975, *Abortion and the Sanctity of Human Life: A Philosophical View*, MIT Press, Cambridge.

Callahan, D.: 1970, *Abortion: Law, Choice and Morality*, MacMillan, New York.

Clouser, K.D.: 1973, ' "The sanctity of life": an analysis of a concept', *Annals of Internal Medicine* **78**, 119.

Compassion in Dying *et al.* v. The State of Washington, U.S. District Court, No C94–119R.

Cronin, D.A.: 1958, *The Moral Law in Regard to the Ordinary and Extraordinary Means of Conserving Life*, Pontifical Gregorian University, Rome.

Nancy Beth Cruzan vs. Director of Missouri Department of Health, 111 L.Ed.2d 224, 110 S.Ct.284, 58 U.S.L.W. 4916 (U.S. June 26, 1990).

Cruzan v. Harmon, 760 S.W.2d 408 (MO.Banc 1988).

Didache: 1965, in *The Apostolic Fathers*, K. Lake, (trans.), Harvard University Press, Cambridge, MA., vol. 1.

Engelhardt, Jr., H.T.: 1996, *The Foundations of Bioethics*, (second edition), Oxford University Press, New York.

Fletcher, J.: 1977, 'Our right to die', *Theology Today* **8**, 202–212.

Frankena, W.K.: 1977 'The ethics of respect for life', in O. Temkin, W.K. Frankena and S.H. Kadish (eds.), *Respect for Life*, Johns Hopkins Press, Baltimore.

Fujii, M.: 1991, 'Buddhism and bioethics', in B.A. Brody, B.A. Lustig, H.T. Engelhardt, Jr. and L. McCullough (eds.) *Bioethics Yearbook: Volume 1*, Kluwer Academic Publishers, Dordrecht.

Hillis, N.G.: 1976, *The Better American Lectures*, Better America Lecture Service, New York.

Irenaeus: 1979, 'Against heresies', in A. Roberts and J. Donaldson (eds.), *The Ante-Nicene Fathers*, vol. 1, W. Eerdmans, Grand Rapids, pp. 315–567.

Kaplean, P.: 1981, *To Cherish All Life: A Buddhist view of Animal Slaughter and Meat Eating*, Zen Center Publications, Los Angeles.

Nolan, K.: 1993, 'Buddhism, Zen, and bioethics', in B.A. Lustig, B.A. Brody, H.T. Engelhardt, Jr. and L. McCullough (eds.), *Bioethics Yearbook: Volume 3*, Kluwer Academic Publishers, Dordrecht.

Noonan, J.: 1970, 'An almost absolute value in history', *The Morality of Abortion*, Harvard University Press, Cambridge.

Payton, S.: 1992, 'The concept of the person in the *parens patria* jurisdiction over previously competent persons', *The Journal of Medicine and Philosophy* **17**, 605–645.

Pope Pius XII: 1958, 'The prolongation of life', *The Pope Speaks* **4**, 393–398.

Rahner, K.: 1968, *Spirit in the World*, Herder and Herder, New York.

Ramsey, P.: 1967, 'The sanctity of life', *The Dublin Review* **241**, 3–23.

Ratanakul, P.: 1988, 'Bioethics in Thailand: The struggle for Buddhist solutions', *The Journal of Medicine and Philosophy* **13**, 301–312.

Rinpoche, S.: 1992, *The Tibetan Book of Living and Dying*, Harper, San Francisco.

Schotsmans, P.: 1991, 'When the dying person looks me in the face: The ethics of responsibility for dealing with the problem of the patient in a persistently vegetative state,' in K. Wildes, F. Abel, J.C. Harvey (eds.), *Birth, Suffering, and Death: Catholic Perspectives at the Edges of Life*, Kluwer Academic Publishers, Dordrecht.

Stevens, J.: 1990, *Lust for Enlightenment*, Shambhata Publications, Inc., Boston.

St. John-Stevas, N.: 1964, *The Right to Life*, Holt, Rinehart and Winston, New York.

Smith, H.L. 1970, *Ethics and the New Medicine*, Abingdon Press, Nashville.

Wildes, K. Wm.: 1991, 'Life as a good and our obligation to persistently vegetative patients', in K. Wm. Wildes, F. Abel, J.C. Harvey (eds.), *Birth, Suffering, and Death: Catholic Perspectives at the Edges of Life*, Kluwer Academic Publishers, Dordrecht.

Williams, G.T.: 1957, *The Sanctity of Life and the Criminal Law*, Knopf, New York.

Wojtyla, K.: 1979, *The Acting Person*, D. Reidel Publishing Company, Dordrecht.

Young, D.R.: 1932, *The Sanctity of Life: Secular and Mid-week Services*, Epworth Press, London.

EDWARD W. KEYSERLINGK

QUALITY OF LIFE DECISIONS AND THE HOPELESSLY ILL PATIENT: THE PHYSICIAN AS MORAL AGENT AND TRUTH TELLER

I. INTRODUCTION

The subject of this paper is that of quality of life and patients with incurable diseases. The paper has four related goals. One goal is to discuss the meaning and contribution of the term "quality of life," and how it relates historically and conceptually to a criterion which preceded it, and one which followed it. The earlier guideline or approach was that of "ordinary verses extraordinary" treatment. Little is heard or seen of this rubric today. The currently favored guideline or approach tends to be that of "medical futility," although "quality of life" also remains alive and well. I will illustrate that, although the labels have evolved, the focus of all three approaches is very similar. I will also suggest that quality of life considerations constitute the essence of one of the meanings of "medical futility."

A second aim is to show how each of these approaches has been and still is used in an attempt to achieve an impossible dream (or better, "delusion"), namely, complete objectivity or neutrality in medical decision-making. The important and lasting contribution of these approaches has been to focus our attention on the right questions to be asked. They should not, however, be viewed as pre-packaged, pre-cooked, pre-digested formulas, as excuses for evading medical judgment or escaping value conflicts.

A third goal is to present a case in favor of physicians as moral agents and tellers of the whole truth in the way they handle quality of life discussions regarding hopelessly ill patients. I conclude that truth-telling means that the physician should tell the whole truth, not simply the so-called objective medical facts, but also his or her value-judgments and recommendations. Some claim that the only legitimate role of physicians is that of establishing and communicating to patients,

Kazumasa Hoshino (ed.), Japanese and Western Bioethics, 103–116.

families and other surrogates objective, value-free medical facts. That view reduces the physician to the role of mere mechanic, it rejects the traditional nature of medicine which is art as well as science, and it pretends that there are such things as value-free medical facts.

A fourth objective has to do with the role and place of the family in these quality of life or medical futility decisions. In advocating for physicians to be moral agents, I do not claim that patients, families and other surrogates don't have primary roles as moral agents. Rather, I concentrate on the role of the physician, but in dialogue, in relationship with the patient's family. I will focus more on the patient's family than on the patient. The competent patient is of course the prime moral agent, but the role of the family speaking for the now incompetent patient who is hopelessly ill but not necessarily dying presents the more interesting and challenging problem. The point here will be that the physician may not usurp the family's role, nor lightly disagree with their wishes, nor impose his or her values upon them; however, in making critical care decisions on the basis of quality of life or medical futility, the family's moral agency should be informed by the physician's moral agency, and vice versa.

For the physician to relate to family members (or to the patient) as mere mechanic, as mere scientist, would be to deny them the full value from the physician-patient relationship. Families and patients want, whether in Japan, the United States or Canada, an account of the medical facts and options and also quality of life evaluations, advice and recommendations made on the basis of the physician's expertise, experience, wisdom and stated values.

II. ORDINARY/EXTRAORDINARY TREATMENT

The approach or criterion which preceded that of "quality of life," was that of the distinction between "ordinary" and "extraordinary" treatment. It was initially proposed by Roman Catholic moral theologians and subsequently widely adopted in medical ethics, in clinical practice and by the courts. But the intended insights were very quickly misunderstood by some. For them it became an instrument for judgment-avoidance rather than one inviting discernment and judgment as intended. This happened in two ways.

Some concluded wrongly that a number of interventions are always by definition either ordinary or extraordinary. In fact the original analysis maintained exactly the opposite, that whether an intervention is ordinary, that is, to be provided, depends entirely upon the patient's status and prospects, and in the final analysis upon how the patient evaluates the burdens and benefits associated with the available interventions. There was a presumed medical objectivity in the ordinary/extraordinary distinction, as compared to the subjectivity inherent in quality of life criteria. Paul Ramsey, for example, wrote that "[t]he terms ordinary-extraordinary, however cumbersome, opaque and unilluminating, directed the attention ... to objective considerations in the patient's condition and in the armamentarium of medicine's remedies" (1977, p. 44).

The second way the original intent of this distinction has become misused to evade judgment and debate is by restricting its focus almost exclusively to dying patients. In fact the original and major breakthrough provided by this analysis was the realization that even a patient who is not dying is entitled to decide that the available treatments are too burdensome and may be refused even if death is a result of that refusal. It was taken as evident by earlier ethicists and physicians that medically prolonging life when death is imminent and unavoidable would be "extraordinary" and not required (Keyserlingk, 1985c). That being so, patients in advance, or families and other surrogates faced with deciding to continue or stop medical feeding and hydration for a permanent vegetative state (PVS) patient, may decide that life support is a burden without benefit, and has become extraordinary.

III. QUALITY OF LIFE

The next approach to appear on the scene was one focused on the patient's "quality of life." Moral calculations on that basis are made by patients about their own lives, or, if they are not able to do so, by families, physicians and others reflecting about that particular patient. Quality of life criteria are captured in the following question. Assuming that at least biological life can be maintained for a brief or long period of time, and in view of available treatment and care and their outcome, is it worth maintaining life according to the patient's own calculation, or own values or best interests? The response is evaluated in view of a cluster of attributes, states, functions or capacities. Among the dimensions and activities of life typically evaluated are the following: the physical, for instance, medical diagnosis and prognosis, symptoms, mobility, and self care; the psychological, for instance, depression, anger, and hope; the social, for instance, relationships, and recreation; the cognitive, for instance, memory, judgment, and alertness; the spiritual, for instance attitudes about the meaning of life and death (Orr, 1993, p. 2). Having done such an evaluation one might conclude that "some qualities of life are not worth living" (Engelhardt, 1975).

The quality of life approach is essentially a more elaborate version of the ordinary/extraordinary distinction. Both involve empirical and moral calculations about whether it is worth maintaining life given the patient's condition and prospects, and the burdens and benefits of available treatment and care. Both assume that calculations will be personal and subjective when patients are the calculators. Both encourage as much "objectivity" as possible when families and physicians are doing the medical and moral estimates, but neither approach expects full objectivity to be possible.

A number of criteria have been proposed for making life and death decisions on the basis of quality of life considerations when the patient cannot do so and others are faced with the decision. Some are applicable to a wide range of medical interventions (President's Commission, 1983), others to specific interventions such as ICU admissions (Engelhardt and Rie, 1986). In Canada and the United States, both law and many ethical analyses now permit the withholding or stopping of life-prolonging treatment in the following circumstances.

(1) Death is irreversible and imminent, and available treatment can only prolong the dying process for a short time.
(2) Whether dying or not, the patient has suffered such extensive brain damage, including the loss of all capacity to relate and communicate with others, that it would not be reasonable to think of him or her as having any interests at all.
(3) Whether irreversibly dying or not, the patient's suffering is extreme and intractable.
(4) The seriously compromised patient's life can only be sustained by excessively burdensome means, or means which could actually worsen the patient's condition (Keyserlingk, 1985, p. 180).

These and similar quality of life criteria obviously provide no support for those who seek objective medical answers precluding the need for value judgments and troubling ethical dilemmas. The criteria just referred to do not consist in empirically verifiable tests. There is indeed a quotient of medical data involved. But deciding, for example, that another's suffering has become extreme and intractable enough to justify stopping life support, requires a moral calculation about the empirical data.

There are two ways in which quality of life considerations have been used to evade moral reflection in favor of an illusionary objectivity. One is by concluding that quality of life factors are inherently and always in conflict with the sanctity of life, and therefore must be rejected as having no normative value. Those advocating that view claim that quality of life criteria are relative, inherently subjective, and about human worth and utility, whereas respect for the sanctity of life requires that all lives be valued equally. Therefore, they claim, no matter how compromised, or how poor the prospects, patients must always be maintained as long as there is biological life, including therefore the PVS patient.

There is, however, no support in biblical, theological, philosophical or legal sources for the conflict between quality of life considerations and respecting the sanctity of that patient's life, as long as the focus is on "value to self," not on utility or worth to others (McCormick, 1978; Keyserlingk, 1979; 1985b). On the contrary, it is precisely because every life is to be respected that we cannot escape grappling with moral questions such as what capacities, what functions constitute a human life worth living for the patient who never expressed a view and has lost the capacity to do so now. Given the new moral ambiguities fostered by the success of medicine, surely the highest form of respect for the sanctity of human personal life is not to lose sight of the distinction between medical success which is merely technological and that which is also human and personal.

A second means of achieving an illusory and inappropriate medical objectivity may well be that of applying to decisions at the bedside quality of life formula designed for health policies about cost-effectiveness. Even in the context of health economics the validity of attaching numbers to functions and illnesses remains the object of much debate, but that is particularly the case when such calculations are made for bedside decisions (Rawles, 1989; LaPuma and Lawlor, 1990). Many assume that quality of life can be accurately and incontrovertibly measured. Assignments of numerical equivalents for functions and disabilities can give the impression of empirical certainty and objectivity. However, they obviously depend upon prior moral judgments about the relative significance or worth of those functions. It is surely a worthwhile enterprise for reasons other than expectations of achieving objectivity, to ascertain community preferences. But those in the community contributing to that data are in effect expressing moral judgments or assumptions in the preferences they express. Furthermore, an unresolved question is what scope to allow for individual preferences which depart from community preferences.

IV. MEDICAL FUTILITY

The third approach attempts to make decisions about critical and terminal care on the basis of whether an intervention is "medically futile" or not. "Medical futility" is without doubt the most popular ethical and legal test in contemporary North American bioethics and law. However, while the expression "quality of life" is heard less, "medical futility" is in fact essentially the same product in a new package.

At the same time, its arrival on the scene does reflect a new stage in North American bioethics, and has occasioned an important debate with serious conceptual and clinical implications. Both because it is an offspring from quality of life thinking, and because it adds a new twist, it merits some attention in this paper.

The origin of decisionmaking on the basis of "medical futility" appears to be a response to a new reality in clinical practice. The earlier issue was essentially that of patients and families trying to get physicians and hospitals to stop un-wanted medical treatment and life support. Their claims, based in large part on the ordinary/extraordinary and quality of life criteria, helped to establish the moral and legal right of patients to refuse treatment. Generally speaking that right is now acknowledged and respected by physicians. The more recent phenomenon is that of patients and families requesting, or insisting, that life-supporting treatment be initiated or continued even when the physician thinks it is no longer appropriate. It does not happen as often as some suggest, but it does happen. The response of those physicians who disagree with the request is typically that the life-supporting treatment requested is "medically futile."

What is meant by "medical futility" and what are the implications of this approach? It has come to have two quite different meanings, and the difference between them has serious implications for the role of the physician in decisionmaking. One meaning is relatively straightforward, namely that the intervention in question simply cannot have the desired effect. For example, on the basis of a scientific calculation of the probabilities, the chances of successfully resuscitating a particular patient, should he arrest, are zero or too low to act upon. In other words, it is futile in the physiological sense. Everyone agrees that it is perfectly appropriate for the physician to make this essentially scientific determination, and that he or she is entitled to refuse to proceed once that futility is soundly established. In the vast majority of cases patients and families, once informed of that futility, do not continue to insist.

But there is a second meaning of "medical futility" used and promoted by many physicians and ethicists, a meaning that in effect incorporates the essence of quality of life thinking. This second sense of futility is concerned not simply with whether a contemplated intervention can be physiologically effective, but with "net benefit." When the intervention can be physiologically effective, (for instance a PVS patient who is capable of being respirated and medically fed and hydrated), the further question is, given the patient's status and prospects, does it benefit the patient to initiate or continue those interventions, does maintaining life under those conditions violate the patient's interests, or some other value such as the ends of medicine? One can immediately see why this question is in essence a "quality of life" question. In practice what is understood by this wider sense of "medical futility" is this question: "Would the use or continuation of these medical interventions in this case be futile, all things considered?" Strictly speaking, therefore, this meaning of "medical futility" is about "the futility or usefulness of medicine."

V. THE PHYSICIAN AS MORAL AGENT

Commentators do not generally deny the legitimacy of this wider "net benefit" meaning of "medical futility," and they grant that patients, or families on their behalf, are entitled to make choices on that basis. However, some object strenuously to physicians having any role in calculations of net benefit or quality of life. They should be restricted, it is claimed, exclusively to conveying in a neutral and detached manner the scientific, medical facts. They also maintain that physicians may not normally refuse a request by a patient or patient's family for a life-prolonging intervention which can be physiologically effective (for instance respirator support and tube feeding for a PVS patient), even though the physician feels strongly that it will not provide a net benefit. Consider, for example, the following stance.

This approach to defining futility replaces a medical assessment (i.e., whether a reasonable potential exists for restoring cardiopulmonary function to the patient) with a nonmedical value judgment that is made by the treating physician (i.e., whether 1 day, 1 week, or 1 month of survival by the patient – perhaps in a severely debilitated state – is of value to him or her (Council on Ethical and Judicial Affairs, AMA, 1991, p. 1870).

Veatch and Spicer (1992, pp. 20, 28) expressed a similar view when they claimed the following:

. . . medical professionals supported by adequate medical science will be presumed to be experts on the medical facts . . . medical professionals can claim no special expertise in deciding the value of outcomes . . . Recall, first, that while we may assume, at least provisionally, that clinicians are skilled in knowing which care will have a relevant effect, they in no way have special skill in knowing whether that effect will be good or bad on balance.

In effect these are claims that the only futility issue which falls within the physician's role and expertise is that of mere physiological futility. In doing so they assume a dichotomy between the physician as scientist and the physician as moral agent, and in effect that the proper and exclusive commitment of medicine is to scientific objectivity and certainty.

Both assumptions are without strong roots or support in the history of medicine, ethics or law. Medicine is a moral as well as scientific enterprise. It is art as well as science. Medical students have always been taught that it is of the very essence of medicine to reflect upon when an intervention is beneficial from moral as well as scientific perspectives. The physician as mere mechanic or technician is no less foreign to humane medicine than the "doctor knows best" variety. Of course the patient and patient's family are the primary decisionmakers, and of course their choices and values should normally decide the issue and should only by exception be challenged or refused, and if so in appropriate ways. Of course patients and families, even in Japan, do not have uniform views on these contentious matters (Bai, 1987, p. 18). However, none of these realities justify the view that physicians do not and should not have moral as well as scientific expertise. Both should play a role in their relationship with patients and families.

Regrettably, critics of the moral agency of physicians presume there are only two choices for physicians in this regard. At one extreme they posit the physician who paternalistically imposes his or her values on the patient or family by refusing their requests or demands for treatment that the physician feels is of no benefit. In North America this paternalistic extreme is far less prevalent than it once was. In Japan on the other hand, where physicians continue to have considerable power and prestige and where patients tend to defer to their authority and power (Feldman, 1985, p. 21), practice at this extreme appears to be very prevalent. At the other end of the spectrum is the physician who simply determines the values and wishes of the patient or patient's family, uncritically offering them items from the treatment menu to match their tastes, and serves it up as ordered.

Compared to clinical realities as they really are, this cast of characters at both extremes are mere one dimensional caricatures rather than real people. Patients and families of patients in fact very seldom insist upon non-beneficial life-support. When they do, it is often precisely because the physician and other health care professionals did not exercise their moral agency at some earlier point by being more honest, more informative and more prepared to propose and recommend moral evaluations and net benefit calculations about the medical options. At the other extreme, physicians seldom sit quietly and passively, merely describing the medical options and scientific probabilities and serving up the treatment ordered. When they do it may well be due to what Jennings referred to as the "... backlash in the medical community in response to the merely technocratic role physicians have been reduced to by the movement for individual self-sovereignty in dying" (1992, p. 113).

There is, in other words, a third place for physicians as moral agents in decisions about futility and quality of life, a position somewhere between the two extremes of imposing their values on others, or uncritically conceding to the value choices of others. It involves the moral agency of the physician in dialogue with that of the patient or patient's family. It includes persuading and recommending, in part on the basis of the physician's moral reflections on his or her medical experience. It includes the physician's experience and reflections about how other patients and families faced with similar choices made their values calculations about quality of life or futility. It includes sharing with them his or her reflections about the difference between apparent and real benefit, and thoughts about the goals and limits of medicine as well as the physician's personal and professional ethics applied to this case. It could even include the physician's ruminations with them about what he or she would be inclined to do if faced with the same choice, and why.

The justification and importance of this partly scientific, partly moral dialogue, rests on several realities seldom included in this debate. The first is that generally speaking this is what patients and the families of patients appear to want. They came to the physician in the hopes of finding both a highly competent scientist combined with an empathetic human being – a tall order, but happily not as rare as some imagine.

They would like some help in evaluating options, in deciding which to choose for themselves or their son, husband, wife, or daughter.

Secondly and similarly, we should not assume, as critics of the physician as moral agent do, that patients or their family members, whether in North America or Japan, have all those value choices neatly sorted out in advance of the medical crisis and decision. That unrealistic expectation appears to lie behind the injunction that the physician's role is simply that of identifying the patient's goals and desired benefits, and acting accordingly (Council on Ethical and Judicial Affairs, AMA, 1991, p. 1870). That is sometimes possible, but not always. In reality this medical crisis, which is forcing a decision, may well be the first time the patient has had to reflect on his or her basic value preferences. The wife of a now incompetent husband may have at best only a vague notion of the values and preferences of her husband relevant (for instance) to whether life-support should continue or stop. They may well need the physician's assistance regarding what values they should think about, what questions they should ask in view of various benefits and outcomes, and how to weigh those outcomes in terms of net benefits.

I hope it is clear that by maintaining that families want and need moral assistance in these decisions, I intend no belittling or down playing of the family and its role. Quite the contrary. It is precisely because a patient's family has such a central role that it deserves the attention and assistance which I am advocating. I sometimes suspect that both those who are quick to reject family choices and those who accept them too uncritically may share a common view that the patient's family is more of an obstacle and nuisance than a legitimate and positive moral force. That negative view of the patient's family is stated or implied in some bioethical writing in North America, but apparently seldom in Japan. What in my view may help Japanese analyses and practice to avoid this danger is the important cultural place occupied by the patient's family, and the tendency to view the patient not as an isolated individual but as an integral member of a family.

Whether in North America or Japan, physicians who too quickly reject family preferences and those who accept them too uncritically would prefer to avoid the time-consuming, often emotional and difficult discussions with families when these decisions are being considered. Happily there is a noticeable trend underway in North American bioethical literature restoring the patient to his various contexts, including that of the family (Hardwig, 1990; Nelson, 1992; Jennings, 1992).

Given the unique vantage point and experience the physician has on medical matters, one should expect that one who is reasonably reflective and senior has at least the makings of a philosophy of life and death of more than passing interest to those who have not seen and thought about these things and must make decisions about them. It does appear to me slightly absurd that many of us seem quite prepared to accept a degree of moral guidance from many who have seen and shared far less of life, death, suffering, human frailty, commitment and courage than have many physicians. As has been noted there may well be a degree of bias in the moral view of many physicians on this and related matters. However, that too does not disqualify physicians from contributing their moral views. It only argues for the importance of self-awareness of one's biases, and the need for physicians, as others, to declare openly the values grounding their positions.

Am I claiming that the moral stances and reasoning of physicians is always correct, that the way every physician calculates a patient's quality of life or net benefit is justifiable? By no means. But then about what other profession could we reasonably make such a claim? The point is, faulty reasoning and dubious net benefit calculations by some physicians hardly justifies restricting physicians exclusively to scientific medical data. By that reckoning we should come to the same conclusion about priests, ministers, clinical ethicists, family counsellors, parents and lawyers.

Some of those who insist that physicians should stick to medical, scientific data, and should remain neutral about values and net benefit, and should make no effort to persuade and recommend, do so in the name of protecting the rights and autonomy of patients and their families. It may well be a stance protective of their rights, but not necessarily of their welfare. In fact, a hard look behind the scenes at the real-life clinical context provides clear evidence that we should often have been clearer, more forceful, more convincing about the lack of net benefit in (for example) maintaining the life of a PVS patient. Because we were not, the family not surprisingly concluded from our neutrality, from the fact that we offered them the option of stopping or continuing, that there must be some net benefit to continuing. Otherwise why would the experts who know and think about these things have offered it?

To expect anxious and confused families to make such a decision all by themselves, without a very empathetic net benefit evaluation and recommendation by the physicians involved is to invite a decision to continue. Because to decide otherwise seems equivalent to saying, "go ahead and kill my father." There is an encouraging trend in hospital practice and policies in North America to discourage the offering of options and choices which in reality are not acceptable options when measured against both meanings of medical futility, including that of net benefit (Montreal General Hospital, 1991).

There is some reason to doubt that patients now medically supported in a PVS condition could possibly have had this condition and its potential longevity in mind when they said to a family member something like, "Don't shorten my life." That is apparently what Helga Wanglie, an 86 year old PVS patient in Minneapolis had told her husband some years earlier (*In re* The Conservatorship of Helga M. Wanglie). To jump from that very general statement to the conclusion that this is what she had in mind seems to be a very big and unjustified leap. Her physicians were convinced that life support offered her no benefit and was inappropriate. Her husband, authorized by the court to make the decision, decided otherwise.

Were the physicians, as some have claimed, out of order in taking a position on the basis of net benefit in this case? Not at all. Their view was not forced upon anyone, and in the end did not prevail. But it was an important side of the issue, contributed significantly to the subsequent debate, and came from a medically and morally informed source.

VI. CONCLUSION

My conclusion has to do with the substantive issue itself of whether being maintained indefinitely in a persistent vegetative state can be considered in any sense a benefit, one consistent with the goals of medicine, whether requested or not by a patient or patient's family. This remains an unsettled and divisive question, and to some extent accounts for the attraction of limiting the discussion to aspects such as due process and autonomy.

However, many substantive questions must be asked and answered, not just by physicians, but by society. Among them I suggest the following. Is it compatible with the aims of humane medicine to maintain human life when it has irrevocably lost the capacity to function in the ways which most clearly define human persons? If it is not, then should society allow individuals to demand it? Is it accurate to claim that while maintaining the life of a PVS patient may not be providing a net benefit, it is not harming the patient?

To hold such a view requires us to limit "harm" to physical and conscious suffering alone. But should we not classify the indignity and assault imposed on such patients as real and significant harms, and not just "metaphorical or symbolic" harms as claimed by Veatch and Spicer (1992, p. 25), and others? After all, the infliction of real indignities do not require awareness of them by the victim – they can for instance be inflicted on dead bodies.

Should we as a society consider reversing the onus of proof in this matter? At present it is expected that those who oppose making life support for the PVS patient available must provide evidence of harm to the patient in doing so. Why not require that advocates for its availability should provide evidence that is beneficial or at least not harmful? Is our reluctance to do so on this and similar matters reflective of the medicalization of life and death in our society, to the point that being life-supported has almost come to be seen as a normal stage and entitlement on the way to death? Perhaps the obstacle here is that we as a society agree with that stance of medicine described by Callahan (1993, p. 34) which

> ... takes as its task only the pursuit of health, or the preservation of a decent quality of life, with death as the accidental result of illnesses and diseases thought to be avoidable and contingent, even though in fact still fatal (Callahan, 1993, p. 34).

Honest differences on this issue will no doubt persist. However, only if physicians and other health care providers in both Japan and North America reflect on these and related matters, and make value judgments they are prepared to articulate and defend, will they be in a position to tell the whole truth to patients and families, and not just the truth about so-called "objective" medical data.

McGill University
Biomedical Ethics Unit
Montreal, Canada

BIBLIOGRAPHY

Bai, K., Shirai, Y. and Ishii, M.: 1987, 'In Japan, consensus has limits', *Hastings Center Report* **17**, 18–20.

Callahan, D.: 1993, 'Pursuing a peaceful death', *Hastings Center Report* **23**, 33–38.

Council on Ethical and Judicial Affairs, American Medical Association: 1991, 'Guidelines for the appropriate use of do-not-resuscitate orders', *Journal of the American Medical Association* **265**, 1868–1871.

Engelhardt, H.T.: 1987, 'Some qualities of life are not worth living', in B. Brody, H.T. Engelhardt (eds.), *Bioethics, Readings & Cases*, Prentice-Hall, New Jersey, pp. 181–184.

Engelhardt, H.T. and Rie, M.: 1986, 'Intensive care units, scarce resources and conflicting principles of justice', *Journal of the American Medical Association* **225**, 1159–1164.

Feldman, E.: 1985, 'Medical ethics the Japanese way', *Hastings Center Report* **15**(5), 21–26.

Hardwig, J.: 1990, 'What about the family?', *Hastings Center Report* **20**, 5–10.

In re The Conservatorship of Helga M. Wanglie, No.PX-91-283 (P. Ct. Minn. Hennepin County June 28, 1991).

Jennings, B.: 1992, 'Last rights: Dying and the limits of self-sovereignty', *In Depth* **2**, 103–118.

Keyserlingk, E.W.: 1979, *Sanctity of Life or Quality of Life in the Context of Ethics, Medicine and Law*, Law Reform Commission of Canada, Ottawa.

Keyserlingk, E.W.: 1985, 'Die Strafbarkeit der Nichtbehandlung von Neugeborenen und Kindern in den Vereinigten Staaten von Amerika', **97** *Z St W*, 178–195.

Keyserlingk, E.W.: 1985b, 'Roots of the belief in the sanctity of life', in B.A. Brody and H.T. Engelhardt (eds.), *Bioethics, Readings & Cases*, Prentice-Hall, New Jersey, pp. 174–177.

Keyserlingk, E.W.: 1985c, 'The right to natural death', *Humane Medicine* **1**, 37–40.

LaPuma, J. and Lawlor, E.F.: 1990, 'Quality-adjusted-life-years; Ethical implications for physicians and policymakers', *Journal of the American Medical Association* **263**, 2917–2921.

McCormick, R.: 1978, 'The quality of life, the sanctity of life', *Hastings Center Report* **8**, 30–36.

Nelson, J.L.: 1992, 'Taking families seriously', *Hastings Center Report* **20**, 5–10.

Orr, R.D.: 1993, 'Quality-of-life is not a dirty word', *Update* **9**, 2–7.

Ramsey, P.: 1977, 'Euthanasia and dying well enough', *Linacre Quarterly* **44**, 40–45.

Rawles, J.: 1989, 'Castigating QALYs', *Journal of Medical Ethics* **15**, 143–147.

Veatch, R.M. and Spicer, C.M. : 1992, 'Medically futile care: The role of the physician in setting limits', *American Journal of Law & Medicine* **18**, 15–36.

PART THREE

DEATH, LIFE, AND WELL-BEING

ROBERT M. VEATCH

AUTONOMY AND COMMUNITARIANISM: THE ETHICS OF TERMINAL CARE IN CROSS-CULTURAL PERSPECTIVE

I. INTRODUCTION

While it is sometimes said that nothing changes in matters of ethics, anyone who would say that has failed to grasp the dramatic changes in the ethics of the care of dying patients over the past two decades. In North America literally hundreds of cases have been debated in courts, in hospital ethics committees, and in scholarly articles (Keyserlingk, 1979; President's Commission for the Study of Ethical Problems in Medicine and Biomedical and Behavioral Research, 1983; Cantor, 1987; The Hastings Center, 1987; Veatch, 1989). In Japan debates over the rights of patients to be informed of a terminal diagnosis have stimulated controversy that has reached the news magazines and scholarly articles of the Western hemisphere (Hiatt, 1989; Kimura, 1994). In Europe one country is still suffering the controversies of unconsenting euthanasia (Lifton, 1986; Koch, 1991). Another prides itself in being the first country of the world to develop a formal public policy condoning intentional mercy killing of the terminally ill upon request (Gomez, 1991; Van Der Mass, 1991; DeWachter; 1989).

The Kennedy Institute of Ethics at Georgetown University in conjunction with the Bochum Zentrum für Medizinische Ethik and the Bioethics Program of Waseda University in Tokyo is currently conducting a comparative study of advance directives and surrogate decisionmaking in Japan, Germany, and the United States.[1] It is our hope to understand the ethical and philosophical underpinnings of decisions for the care of the dying in the three cultures.

Kazumasa Hoshino (ed.), Japanese and Western Bioethics, 119–129.

II. INDIVIDUALISM AND ITS LIMITS

The movement to modify the care of the dying in Western culture has been given a great deal of attention in the United States. In addition to the hundreds of legal cases that have been adjudicated in the courts, thousands more have been debated in hospital ethics committees, academic classrooms, and the lay press (Veatch, 1989; Cantor, 1987; Choice in Dying). Increasingly, a consensus is emerging over the underlying moral and legal principles for resolving these cases. When there is agreement that the patient is competent to make his or her own medical decisions and the treatment is proposed for the patient's own good, the autonomous choice of the individual patient to refuse medical treatment has in every single case been found to take precedence over the desire to prolong the patient's life (Veatch, 1989; Meisel, 1992).

Individual self-determination is given extremely high priority in Western culture in general and in the United States in particular. If the case can be formulated as one involving the right of the individual to make his or her own medical decisions and no significant interests of other parties are perceived to be at stake, then the principle of autonomy will prevail, at least if the case reaches the courts. This is not to dispute Tom Beauchamp's critique of the received view presented elsewhere in this volume. Western physicians have for centuries survived as an island of paternalism in the sea of liberalism just as a patients' rights movement exists in Japan in a broader, more traditional culture.

The Religious Roots of Individualism

Western culture emerged gradually, resulting from the integration of the Graeco-Roman culture with that of Judeo-Christianity when Constantine became the first Christian Roman emperor in 312AD. His project was to unite the Christian Church and the secular state. Neither Graeco-Roman nor early Judeo-Christian thought had anything resembling a full-blown affirmation of the individual as decisionmaker. Christianity, however, as a religion of conversion, planted the seeds of the notion of the individual as decisionmaker. In fact, one of the well-known Biblical texts of the tradition praises one who would separate himself from his family for purposes of accepting Christianity. The individual as decisionmaker is on her way to becoming differentiated from both family and cultural group.

The emergence of sectarian and mystical thought of the late middle ages and the Protestant Reformation extended further the affirmation of the individual as having sole responsibility for decisions about life. Mysticism established a direct epistemological link between the individual and the deity. Protestantism affirmed the capacity of the individual to understand the Biblical text and to have authority in interpreting it. While individual self-determination in matters of morality was not yet established, its foundations were.

Secular Western Individualism

The full emergence of the autonomous self-determining individual as an ideal in Western culture occurs with the eighteenth century and the Enlightenment. The secularization of society and the emergence of the rational, free-thinking individual separate from church authority is seen in the philosophical and political writings of Hume (1711–1776), Rousseau (1712–1778), and Kant (1724–1804) and to some extent in their seventeenth century precursors: Locke (1632–1704) and Hobbes (1588–1679). It is not an accident that these intellectual giants of the modern Western tradition were nurtured in the thought of Protestant thinkers such as Luther (for Kant) and Calvin (for Hume).

Jefferson, Madison, and Hamilton are often referred to as the founding fathers of the United States. They crafted not only the Declaration of Independence and the U.S. Constitution, but articulated more generally the spirit of liberalism appropriate for a frontier culture, incorporated the language of rights into the American ethos, and solidified the importance of the individual as decisionmaker into the American ideal. The frontier not only separated people from their family heritage, but also taught the mystique of the rugged individualist.

Terminal Illness Law

This cultural and political philosophy is usually referred to as liberalism. It stresses the freedom and intellectual authority of the individual. Each individual is encouraged to make his or her own choices grounded in a personally formulated life plan. This ideology has become the foundation of the ethical and legal approach to the care of the terminally ill. The ethic of liberal political philosophy was destined to clash with the traditional Hippocratic ethic of Western medicine. That ethic had its roots in Greek medicine, which valued medical authority and secrecy as much as the traditional medical ethics of tradtional Japanese culture. As would be appropriate for a movement growing out of Greek philosophy and mystery religion, there was no hint of a view of the patient as an individual self-determining decision maker. In fact, the Hippocratic Oath requires the physician to keep medical knowledge from the lay person and to treat the patient paternalistically.[2] In this respect the traditional Western medical ethic of the care of the dying was as paternalistic as those of traditional Japanese culture as seen in the "Seventeen Rules of Enjuin," the sixteenth century medical ethical code of the Ri-shu school, which requires keeping medical knowledge secret (Bowers, 1970, p. 8–10).

The visible tension between the liberal political philosophy of the broader culture and the anti-liberal paternalism of the traditional Hippocratic medical ethic did not emerge until the twentieth century. Early in the century cases focusing on the right of patients to give consent to treatment began to emerge, but it was not until the 1970s that the full-blown conflict between these two ethics captured the public attention. By then the capacity of medicine to prolong the lives of critically ill, dying patients suffering from cancer, end-stage renal disease, and severe neurological impairments, forced an increasingly rights-oriented public to demand autonomous control over medical decisions. This was based on the claim that being an expert on the technical aspects of medicine did not give one special authority on the moral or ethical aspects. Especially for the terminally ill, the patients' rights movement was born. The law, a chief repository of the tradition of liberal political philosophy, became the advocate for the self-determination of mentally competent adult patients.

Two legal foundations provided this support: a common law tradition of affirmation of individual self-determination and a constitutional law that was beginning to articulate a doctrine at the time referred to as "privacy." The common law in the Anglo-American tradition is expressed in the sum of judges' opinions in legal cases. Now, as has been indicated, every single judge's opinion (except a few opinions from lower courts that would later be overturned by courts of appeals) has recognized the right of the competent patient to refuse life-prolonging medical treatment if that treatment is offered for the patient's own good (Meisel, 1992; Veatch, 1989). The Constitutional law basis of protection of patients' rights to self-determination is somewhat less firmly established. There is no explicit right of self-determination in the American Constitution. The courts, however, have in recent years articulated a right of "privacy," which they have constructed from several amendments to the U.S. Constitution (Davis, 1978; McGinley, 1980). This reasoning was particularly apparent in the early legal cases such as that of Karen Quinlan (In Re Quinlan, 1985). She was a twenty-one-year old woman left in a persistent vegetative state following what was believed to be an overdose of a tranquilizing drug combined with alcohol. The state Supreme Court found that she had a right of privacy to forgo life-prolonging ventilator life support and that that right could be exercised by her father on her behalf after she had become unconscious. Similar decisions have been reached in all other similar cases since then, except that recently there has been a reluctance to continue using the word *privacy* when self-determination is really what is meant (Cruzan, 1990).

Challenge to the Liberal Consensus

Until very recently all of the controversy in the care of the terminally ill in the United States could be linked to cases in which autonomy and self-determination could not be relied on to settle the decisionmaking. They were invariably cases in which either the patient was not clearly competent or the patient's decision jeopardized the welfare of other people.

1. *Incompetent Patients: Autonomy Extended*

If the patient is not competent to make his or her own decisions, clearly no principle of individual autonomy will be relevant. The one exception is in the case of the formerly competent person who has left clear instructions regarding his or her wishes. The American discussion now recognizes what could be called the *principle of autonomy extended* (Veatch, 1989). If what the patient desired while competent can be determined, that desire will prevail over the views of either physician or family (Cruzan, 1990). The individual retains priority over family or community even after lapsing into incompetency. Only very recently has this notion been called into question and then only with regard to patients who are permanently disabled to the extent that they cannot recall their former selves and the wishes that were expressed (Buchanan and Brock, 1989; Dresser and Robertson, 1989).

2. *Limited Familial Autonomy*
If the patient's wishes cannot be determined, then increasingly the family is perceived as the appropriate surrogate (Areen, 1987; Veatch, 1984). A case can be made that the family is given discretion under the rubric of what I call *limited familial autonomy*. Under this notion the family is viewed as an autonomous social unit analogous to the autonomous individual. This would provide a basis for the family to make choices based on the family's values when (and only when) the individual's values are not known. In the U.S. the familial authority is clearly second ranked in such a scheme. As in the case of the formerly competent patient, the never-competent patient presents a problem that has been addressed in the American culture by trying to convert it into a problem of autonomy. In this case familial autonomy becomes a surrogate for the autonomy of the individual.

3. *The Harm-to-Others Principle*
In rare circumstances the principle of autonomy presents a problem because the autonomous choice of the patient will produce harms to other people, either specific people or society in general. The most common cases are those in which a terminally or critically ill person's refusal of life-supporting medical treatment will lead to death, which, in turn, will harm the interests of someone dependent on the refuser of care. For example, if Jehovah's Witness parents refuse life-prolonging blood transfusions, their refusal will lead to abandonment of their children. In such cases, treatments have been ordered against the wishes of the competent adult (Raleigh-Fitkin, 1964). This is consistent with the harm-to-others principle, which recognizes a limit on individual autonomy in cases in which an individual's decision will produce significant harm to others. There is some dispute about whether it is the amount of harm that is relevant or the moral relationship between the refuser and the one harmed. Generally, it is recognized that there must be some on-going commitment (such as in a parent-child relationship) to justify forced treatment. Thus, some theorists would claim that contemporary Anglo-American bioethics requires a correction of the general principle that harm to others justifies violation of autonomy. In the newer and better understanding, only in cases in which specific duties to others require overturning autonomy – duties grounded in promises or in justice – would such an overturning be acceptable (Veatch, 1981).

Recently, there has been dispute over the role of autonomy in permitting physicians to refuse to provide life-sustaining medical care desired by a competent individual either for herself (Angell, 1991) or for a dependent (In the Matter of Baby K, 1993). It is a misunderstanding of the ethical principle of autonomy to believe that autonomy would give such patients a right of access. Autonomy generates only a "negative right," that is the right to be left alone to pursue one's goals with others willing to provide assistance; it does not give one a right of access. In fact, insofar as providing such care produces harm to others, the harm principle would generally limit access especially when the autonomy of health professionals is also violated. If people have a right to care that will temporarily extend life of an unconscious patient and force physicians to cooperate in that project, their right must be founded on some other ethical consideration besides autonomy. It must rest on a previous promise made or on some general claim of justice, but not on autonomy (Veatch and Spicer, 1992). Here I differ from Kyserlingk in holding that licensed physicians have made certain promises to deliver care and that those promises should be kept even if the physician's personal value judgment is that there is no net benefit in the treatment.

The Challenge to Individualism

Only very recently in American culture has anyone begun to question the presumption in favor of individual self-determination in cases in which a patient is competent and the patient's decision does not involve a serious burden to others. A movement known as *communitarianism* has emerged that has questioned the ultra-individualism of American liberalism (Emanuel, 1991; MacIntyre, 1981; Callahan, 1990). Holders of this view press beyond the mere affirmation of the right of the individual to choose to a recognition that individuals exist within communities, and their decisions about terminal care should be shaped by their membership in such communities. This has led Daniel Callahan, for example, to press for the acknowledgement that there comes a time when one has had a full life and that it is time to step aside to let others have access to health care needed to complete their lives. The individual is seen as part of a larger community of interests in which the common good has moral priority over the wishes of the individual, a notion probably more familiar to those accustomed to Japanese culture than to American individualism.

It is too early to tell whether this movement is merely a modest delay in the progression toward an ever more individualistic liberal society. In that society life-prolongation is made available on the basis of the individual's personal decisions within the framework of ability to pay. Alternatively, this may be the first stages in the abandonment of liberal individualism and a movement more in the direction of seeing a moral community as bound together in families, neighborhoods, and cultural identity. It seems clear that the interests of some other parties – family members, colleagues, and certain members of the broader community – have returned to the scene in the ethics of terminal care in the United States. As such we may be moving in the direction of the Japanese understanding of *amae* (Doi, 1981), of dependence or "bondedness," which more closely resembled ancient Western culture, but has been lost to the individualism of the modern West.

III. JAPAN: THE CHALLENGE OF INDIVIDUALISM

It is striking that just as American liberal individualistic culture is challenged by communitarianism, feminist theory, and other anti-individualistic movements, Japan appears to be showing much greater interest in individual rights, the doctrine of informed consent, and the trappings of individual patient self-determination. Reports that have reached the West have signaled that Japan's bioethics movement is becoming more interested in the right of the individual patient to know his or her diagnosis, give informed consent or refusal of consent to treatment, and even to prepare advance directives (Kimura, 1994; Ninomiya, 1978; Hattori *et al.*, 1991; Kai *et al.*, 1993). The work of Tomoaki Tsuchida has been particularly insightful in revealing how uniquely Western the presumption of individualism is and how it is only lately intruding on traditional Japanese culture. Contrasting Japanese cultural response to death with that of the individualist liberal West, Tsuchida says:

> [f]or the American, it is not only a right to exercise control over one's own destiny, but also one's duty. Death and life are one's own private concern. The Japanese, by contrast, have lived for centuries in a highly integrated and contextualized society where even life and death have to be seen as a family affair – it not the affair of the community as a whole – as much as the affair of the particular individual. Without the consent of the family, a doctor is not expected to inform a patient of a fatal illness or even to undertake serious surgery, much less organ transplants (Tsuchida, 1992, p. 321).

But at a meeting of the project on advance directives and surrogate decisionmaking in the spring of 1994, Professor Tsuchida made clear that this traditional Japanese integration of the family and the community is being challenged by a number of influences including the breakdown of the Japanese family and Western ideological influences such as the emphasis on the rights of patients as individual decision makers.

IV. CONCLUSION

There is an overwhelming tendency to hold out a "golden mean" as the proper resolution to a dichotomy in which two cultures are each seen has converging from ideal typical patterns. In this case one sees the United States as the example of a culture that has been hyper-individualistic and liberal in its emphasis on patient autonomy in decisionmaking for the terminally ill while Japan was more traditionally communitarian. It is tempting to conclude that each represented an extreme for which a corrective was needed and that movement toward some more balanced harmony of the two patterns is what should emerge. This view of an "entropy of cultures" may have some wisdom in it, but this is not necessarily so. It may be that there is value in diversity so that it would be just as much a tragedy for the Japanese for the traditional communitarian patterns of Japanese culture to be lost as it would be a tragedy for individualist, egalitarian Americans to be forced away from the richness of its ethic of democratic liberalism back to the paternalism of an outdated Hippocratic ethic. There is great good to be gained from a mutual understanding of each culture's way of making terminal illness decisions. Whether that should also lead to a compromise between the two patterns remains to be seen. Perhaps the wiser approach would be to search for the core universal ethical principles while encouraging the flourishing of a plurality of cultural variations in the specifics.

Georgetown University
Washington, D.C., U.S.A.

NOTES

[1] Chief Investigators are Hans-Martin Sass, Director of the European Professional Ethics Program, Kennedy Institute of Ethics, and Member of the Zentrum für Medizinische Ethik and Robert M. Veatch, Director of the Kennedy Institute of Ethics, assisted by Rihito Kimura, Director of the Asian Bioethics Program, Kennedy Institute

of Ethics and Professor of Law and Bioethics at Waseda University, Tokyo. The project is funded by the Volkswagen Foundation. See Sass, Hans-Martin, Robert M. Veatch, and Rihito Kimura, *Advance Directives and Surrogate Decision-making: Cross-Cultural Perspectives*, forthcoming.

[2] This was consistent with the Pythagorean view of knowledge as dangerous and follows the pattern of other Greek religious cults that held it as a high duty to keep the mysteries of the cult secret from the uninitiated masses. Edelstein interprets the pledge that the physician will protect the patient from harm as a paternalistic injunction to protect the patient from himself. See Edelstein, Ludwig: 1967, 'The Hippocratic oath: Text, translation and interpretation,' *Ancient Medicine: Selected Papers of Ludwig Edelstein*, Temkin, Owsei, and C. Lilian Temkin, (eds.), The Johns Hopkins Press, Baltimore, pp. 3–64.

BIBLIOGRAPHY

Angell, M.: 1991, 'The case of Helga Wanglie: A new kind of "right to die" case', *The New England Journal of Medicine* **325**, 511–512.

Areen, J.: 1987, 'The legal status of consent obtained from families of adult patients to withhold or withdraw treatment', *Journal of the American Medical Association* **258**, 229–235.

Bowers, J.Z.: 1970, *Western Medical Pioneers in Feudal Japan*, The Johns Hopkins University Press, Baltimore.

Buchanan, A.E. and Brock, D.W.: 1989, *Deciding for Others: The Ethics of Surrogate Decision Making*, Cambridge University Press, Cambridge.

Callahan, D.: 1990, *What Kind of Life: The Limits of Medical Progress*, Simon and Schuster, New York.

Cantor, N.: 1987, *Legal Frontiers of Death and Dying*, Indiana University Press, Bloomington.

Choice in Dying, Inc.: *Right to Die Law Digest Statutes*, Choice in Dying, Inc., New York.

Cruzan v. Director, Missouri Dept. of Health, 110 S.Ct. 2841 (1990).

Davis, P.K.: 1978, 'Constitutional law-right of privacy-qualified right to refuse medical treatment may be asserted for incompetent under doctrine of substituted judgment', *Emory Law Journal* **27**, 425–460.

De Wachter, M.A.M.: 1989, 'Active euthanasia in the Netherlands', *Journal of the American Medical Association* **262**, 3316–3319.

Doi, T.: 1981, *The Anatomy of Dependence*, Kodansha International Ltd., Tokyo.

Dresser, R.S. and Robertson, J.A.: 1989, 'Quality of life and non-treatment decisions for incompetent patients: A critique of the orthodox approach', *Law, Medicine, & Health Care* **17**, 234–244.

Emanuel, E.J.: 1991, *The Ends of Human Life: Medical Ethics in a Liberal Polity*, Harvard University Press, Cambridge.

Gomez, C.F.: 1991, 'Euthanasia: Consider the Dutch', *Commonweal* **118**, 469–472.

Hattori, H., Salzberg, S. M., King, W. P. *et al.*: 1993, 'The patient's right to information in Japan: Legal rules and doctors' opinions', *Social Science and Medicine* **32**, 1007–1016.

Hiatt, F.: 1989, 'Japan court ruling backs doctors: Judge says that patients may be kept ignorant of their illnesses', *Washington Post*, May 30, A-9, 16.

In re Quinlan, 70 N.J. 10, 355 A. 2d 647 (1976), *cert. denied* sub nom., Garger v. New Jersey, 429 U.S. 922 (1976), overruled in part, In re Conroy, 98 NJ 321, 486 A.2d 1209 (1985).

In the Matter of Baby K, 1993 WL 343557 (E.D. Va.).

Kai, I. *et al.*: 1993, 'Communication between patients and physicians about terminal care: A survey in Japan', *Social Science and Medicine* **37**, 295–304.

Keyserlingk, E.W.: 1979, *Sanctity of Life or Quality of Life in the Context of Ethics, Medicine and Law*, Law Reform Commission of Canada, Ottawa.

Kimura, R.: 1994, 'Bioethics and Japanese health care', *Washington-Japan Journal* **2**, 1–7.

Koch, H.: 1991, 'Bundesrepublik Deutschland', in A. Eser and H.-G. Koch (eds.), *Materialien zur Sterbehilfe: Eine internationale Dokumentation*, Max-Planck-Institut fur auslandisches und internationales Strafrecht, Freiburg im Breisgau, pp. 31–193.

Lifton, R.J.: 1986, *Nazi Doctors*, Basic Books, New York.

MacIntyre, A.: 1981, *After Virtue*, University of Notre Dame Press, Notre Dame.

McGinley, T.: 1980, 'The right of privacy and the terminally-ill patient: Establishing the "right-to-die" ', *Mercer Law Review* **31**, 603–615.

Meisel, A.: 1992, 'The legal consensus about forgoing life-sustaining treatment: Its status and its prospects', *Kennedy Institute of Ethics Journal* **2**, 309–345.

Ninomiya, R.: 1978, 'Contemporary Japan: Medical ethics and legal medicine', in W.T. Reich, (ed.), *Encyclopedia of Bioethics*, The Free Press, vol. 3, New York, pp. 926–929.

President's Commission for the Study of Ethical Problems in Medicine and Biomedical and Behavioral Research: 1983, *Deciding to Forego Life-Sustaining Treatment: Ethical, Medical, and Legal Issues in Treatment Decisions*, U.S. Government Printing Office, Washington, D.C.

Raleigh Fitkin-Paul Morgan Memorial Hospital v. Anderson, 42 N.J. 421, 201 A. 2d 537 (1964) cert. denied, 337 U.S. 985 (1964).

The Hastings Center: 1987, *Guidelines on the Termination of Life-Sustaining Treatment and the Care of the Dying*, The Hastings Center, Briarcliff Manor, New York.

Tsuchida, T.: 1992, 'From ethos to ethics: Japanese views on life and death', in Institute of Medical Humanities, (eds.), *Toward a New Replenishment of Medical Education and Hospital Service*, Kitasato University, Tokyo, pp. 319–325.

Van Der Mass, P.J., Van Delden, J.J.M., Pijnenborg, L. and Looman, C.W.N.: 1991, 'Euthanasia and other medical decisions concerning the end of life', *The Lancet* **338**, 669–674.

Veatch, R.M.: 1989, *Death, Dying, and the Biological Revolution*, revised edition, Yale University Press, New Haven.

Veatch, R.M. and Spicer, C.: 1992, 'Limits of guardian treatment refusal: A reasonableness standard', *American Journal of Law and Medicine* **9**, 427–468.

Veatch, R.M.: 1981, *A Theory of Medical Ethics*, Basic Books, New York.

Veatch, R.M. and Spicer, C.M.: 1984, 'Medically futile care: The role of the physician in setting limits', *American Journal of Law & Medicine* **18**, 15–36.

FUMIO YAMAZAKI

A THOUGHT ON TERMINAL CARE IN JAPAN

I. INTRODUCTION

Dr. Robert Veatch's article, 'Autonomy and communitarianism: The ethics of terminal care in cross-cultural perspective,' briefs on the history of terminal care in the U.S., which has taken shape within an establishment of individualism. He points out that today's terminal care ethics in the U.S. is the result of thorough disputes and discussions in courts, ethical committees and scholarly articles. I am particularly interested in his observation that recently a concept of communitarianism has emerged over against the philosophy of individualism. Is this because there have been some problems with the individualistic approach to terminal care?

However, if there are problems with the individualism-based terminal care, it is not obvious that the Japanese approach to terminal care should be considered right, in which traditional familialism and communitarianism are still dominant and patients are rarely informed of their disease or diagnosis.

Dr. Veatch concluded that there is a strong tendency both in America and Japan to seek the golden mean as the best approach to terminal care, and that entropy of cultures may have some wisdom in it, but this does not necessarily hold. I think his view is right, considering the present situation of Japanese terminal care ethics. Japanese terminal care is in so miserable a state that it has not even reached the stage in which it can be usefully compared with the U.S. approach.

II. X UNIVERSITY HOSPITAL'S CASE

An incident at X University Hospital, the so-called "Euthanasia Incident," clearly shows the problems of Japanese terminal care. This incident is such that a doctor took the life of a multiple myeloma patient

Kazumasa Hoshino (ed.), Japanese and Western Bioethics, 131–134.

by injecting potassium chloride into his vein. At the time of this writing, this case is still pending in court and the judicial authorities have not yet reached any conclusion. One of the key issues that they are trying to determine is if this act can be considered euthanasia. However, in my view, it is clear that this is not euthanasia. The patient was in a coma when this incident happened. Hence, it is unlikely that he had unbearable physical pain. So, it cannot be said that the doctor freed the patient from pain. Therefore, the question here is not whether this was euthanasia, but why this happened. I would like to point out some of the problems.

Could the Patient's Pain Not Be Removed?

The patient's family, unable to bear to look at his suffering, reportedly asked the doctor to put him at peace. But this poses the question of whether it was really impossible to remove his pain before he went into the coma. For at hospices and PCUs(palliative care units), patients rarely, if ever, thrash about with pain before death. That is to say, there is room to suspect that pain alleviation treatment was not properly conducted. There have been many cases, including this incident, in the Japanese medical community in which terminally ill patients writhe in treatable pain. It is apparently unethical that patients are left in pain because of doctors' ignorance, when there are ways to alleviate it.

Was a Team Approach Taken?

Secondly, the doctor used potassium chloride at his own discretion. It is a grave problem that the decision that affected this human life was made by one single doctor. He was not the only doctor on an isolated island, but one of the staff of a medical team. If the team doctors had had thorough discussions in case conferences, this incident might not have happened. I wonder also how often doctors held joint meetings with nurses. Had they treated this patient, incorporating the opinions of nurses who are closer to patients and their families, this incident would have been prevented. Some might think that my view is unrealistic. I have to admit that it is not a common view in the Japanese medical community, where doctors and nurses face patients separately and seldom have joint meetings. Even if nurses have views that suit the patients' needs, more often than not doctors make final decisions.

However, patients' needs can be met only when doctors and nurses work together. I think that the team approach, where doctors and nurses face the patients with a common understanding, must be the basis for terminal care. The current state of Japanese terminal care, in which the team approach is not yet established and doctors' opinions take precedence to those of nurses, is in itself unethical. The X University incident occurred in part due to such an imbalance of power between doctors and nurses.

Was the Patient Given a Chance to Provide Informed Consent?

The most basic and yet important issue in this case is that the patient was not informed of his disease or of his condition at all. This patient is said to have been given intensive treatment until he lapsed into a coma. If he had been informed of his disease and its terminal condition based upon the doctrine of informed consent, he could have expressed his wishes at each stage of treatment and spent the last days of his life in content. He could have rejected treatment that would cause a great deal of pain. However, it was the doctor and his family who actually made the decision. Moreover, the doctor chose the initial treatment that merely prolonged his life. This is also often the case with Japanese medical practices. Although it is true that in recent years the doctrine of informed consent has begun to be introduced into Japanese medical circles, it is not yet satisfactorily prevalent. For instance, patients with early-stage cancer are often informed of their diseases and conditions and their requests are incorporated in the treatment, whereas patients with progressive or terminal cancer are rarely given such information and are forced to spend their remaining days in lies and empty encouragements. Patients who are not given explanations on their actual condition despite their wishes, are unable to pursue the life they want to lead. This immature state of informed consent is, I think, the most serious ethical problem in Japanese terminal care.

III. THE PRESENT STATE OF JAPANESE TERMINAL CARE ETHICS

After the X University incident, the university's ethical committee criticized the doctor as unethical, which I think is the right view. However, the committee made few comments on the points that I mention here; that is, whether the symptoms were properly controlled, whether the team approach as a basis for the terminal care was taken, and whether the doctrine of informed consent was exercised. Probably the committee had not realized the unethical nature of the terminal care of its own university hospital. This is not a problem with X University Hospital alone, but with almost all the Japanese hospitals, except hospices and PCUs.

VI. CONCLUSION

The ethical problems which Japanese terminal care faces have only begun to be addressed and it will probably take a long time for the results of such studies to permeate into actual medical practices. Therefore, I do not think it is the right time to seek the golden mean between the U.S. approach and the Japanese approach. It is only after Japanese terminal care establishes its own style of individualism-based terminal care ethics, overcoming the problems that have taken root in the communitarianism-based approach through thorough discussions and accumulated experiences, that we should start considering a comparison between the two. We have just begun to take the first steps.

St. John's Hospital
Konganei-shi, Tokyo
Japan

BARUCH BRODY

MEDICAL FUTILITY: PHILOSOPHICAL REFLECTIONS ON DEATH

The recent medical literature in the United States has contained an extensive discussion of the issue of medical futility. Inspired by cases such as that of Helen Wanglie (Angell, 1991), in which the husband of an elderly woman in a persistent vegetative state demanded continued life-prolonging interventions, clinicians and ethicists have debated whether or not patients are entitled to receive life prolonging interventions (demanded by them or the surrogates who speak for them) even when the clinicians judge those interventions to be futile. Some (Blackhall, 1987) have emphasized the objectivity of the clinical judgements that the interventions in question are futile, and have permitted physicians in such cases to unilaterally decide to limit those futile interventions. Others (Youngner, 1988) have emphasized the value-laden nature of such judgements, and have insisted that it is the patient or the family who must decide which interventions will be provided.

My colleague Amir Halevy and I have argued (Brody and Halevy, 1995; Halevy and Brody, 1996) in a series of papers: (1) that there are many different concepts of futility employed in this debate, but that none of them satisfy the conditions which must be satisfied before judgements of futility can be used as the basis for clinicians unilaterally deciding to limit life-prolonging interventions; (2) that the actual incidence in ICU's of futile cases (properly defined) is probably quite small; (3) that decisions to unilaterally limit in a wider set of cases life prolonging interventions might well be justified by considerations of individual/instutional integrity.

In this paper, I want to focus on a different issue. What does the debate about futility teach us about American (and maybe Western) ideas concerning the meaning and value of life and about how Americans (and maybe Westerners) understand the reality of death? In particular, I shall: (a) trace the emergence of the debate about futility from a consensus that

Kazumasa Hoshino (ed.), Japanese and Western Bioethics, 135–144.

had emerged in the U.S. about limiting life-prolonging interventions; (b) show the attitudes of different groups towards life and death before and after the emergence of the debate about futility; (c) suggest that the U.S. (perhaps the West, in general) needs to think more about the meaning of life and death if we are to deal better with the issues raised by the debate. I invite others to suggest ways in which Eastern perspectives on these fundamental metaphysical issues might be of help in dealing with the issue of futility.

I. THE CONSENSUS AND THE EMERGENCE OF THE FUTILITY DEBATE

In the 1970's and 1980's, the following consensus emerged in the United States (and in some other Western countries) about decisions to limit life prolonging interventions.

(1) The mere fact that some treatment exists that would prolong the life of some patient does not by itself suffice to justify providing that therapy to that patient because extending life by providing the treatment in question may not be beneficial to the patient in light of the patient's values and patients have a right to refuse therapy whose provision they judge to be against their interests.

(2) Both competent and incompetent patients have that right to refuse life-prolonging therapy; of necessity, the way in which that right is exercised is different in the two cases.

(3) Despite the differences, decision making for both types of patients should usually occur in the clinical setting without recourse to the courts.

(4) The main role of legislatures is to see that these rights are adequately recognized in law and the main role of health care institutions is to insure that there are proper processes for such decision making and for insuring that they are appropriately documented.

(5) These refusals may be of all forms of therapy, including artificial nutrition and hydration, and they may involve both the withholding of life-preserving therapy and the withdrawing of such therapy.

(6) Such refusals of life prolonging therapy must be distinguished both from active euthanasia and from assisted suicide.

(7) When the patient is not competent, the attending physician, who normally makes that decision of incompetency, may rely upon a patient's advance directive to refuse treatment incorporated in such documents as living wills.

(8) If there is no living will or its equivalent, the patient's right to refuse treatment may be exercised on behalf of the patient by a surrogate decision maker appointed by the patient in advance (by use of such mechanisms as the durable power of attorney) or specified by some statutory scheme (usually, family members in some ordering of priority). The surrogates should apply a substituted judgement standard or, if they cannot, a best interests standard.

(9) The same principles should apply to parental decision making about life prolonging therapy for their children, especially severely ill newborns, but the emphasis must be on the child's best interests since the substituted judgment standard cannot apply.

What is crucial to this consensus is the emphasis on the values and rights of patients. According to the consensus, the value of life prolonging interventions is based upon the values of patients; the very same therapy which keeps two similar patients alive may be beneficial to one (because continued life is beneficial in light of that patient's values) while harmful to the other (because continued life is harmful in light of the other patient's values.) Moreover, according to the consensus, it is the right of patients (or the surrogates who speak for them) to make the decision as to whether or not the interventions in question will be provided.

Those who introduced the concept of futility did not challenge the validity of this consensus in general. They merely argued that there are certain special cases in which we do not need to consider the special values of the patient and the patient has no decision to make; the intervention that just won't work is therefore properly described as futile, and can be limited by the physician's unilateral decision based upon his or her objective knowledge of the futility of the intervention. In fact, there was a debate among those who introduced the concept of futility as to whether the patient (or the patient's surrogate) even needed to be informed about the decision not to provide the futile intervention. We are not required, some argue, to tell people all the things that we aren't doing because we know that they will not work.

Perhaps one example will illustrate the way in which the concept of futility emerged in the context of the earlier consensus about limiting life prolonging interventions. Starting in 1982, the Council on Ethical and Judicial Affairs of the AMA has issued a series of policy statements on ethical issues surrounding the use of do not resuscitate orders and other orders limiting the use of life prolonging interventions. These statements embody most of the elements of the consensus and played a significant role in the development of the consensus. In its 1991 statement (Council, 1991), DNR orders are judged as acceptable either if they represent the patient's wishes or if the resuscitation would be futile. This second reason, introduced for the first time as a supplementation to the traditional autonomy-based reason of the patient's wishes, is defined very narrowly. Resuscitation is futile if it cannot reasonably be expected to restore cardiac or respiratory function or if it will not achieve the patient's expressed goals. The former condition is not often satisfied, while the latter is much closer to the traditional focus on the goals of the patient. The narrowness of the AMA's definition of futility is evidence of the extent to which considerations of futility are meant to supplement, rather than to supplant, the type of decision making to limit life prolonging interventions recognized by the consensus. Other professional groups (American College of Physicians, 1992) have recognized possible extensions of the definition of futility, so that it applies to cases in which survival to discharge cannot reasonably be expected, even if resuscitation restores cardiac or respiratory function. Nevertheless, it is clear that they still view the appeal to futility as a supplement to patient-driven decision to limit care.

II. LIFE AND DEATH BEFORE AND AFTER FUTILITY

What forces called for the development of the initial consensus? Why was a need felt for the development of policy statements by professional groups, legislation (especially living will and durable power of attorney legislation), and hospital protocols that embodied the consensus? A series of famous court cases from Quinlan in 1976 to Cruzan in 1990 suggest an answer to these questions, an answer that sheds much light on certain attitudes to life and death and towards technology.

In most of these famous court cases, the patient (or his or her surrogate) wished to limit the use of life prolonging interventions but the physicians insisted on providing them. The patient (or his or her surrogate) then went to court to secure an order limiting the interventions provided. Why did the physicians in these cases insist on providing the interventions? The records for these cases need to be searched more carefully to answer this question. In the meantime, a number of suggestions can be offered.

One possibility is that the treating physicians thought that the law mandated their provision, and they wanted the family to obtain a court order mandating their discontinuance to protect the physicians. On this account, the physicians really agreed with the family, but wanted protection from the law. This account might have some plausibility in the earlier cases, although even there it faces the difficulty of explaining the physicians' fears given the concurrence of the entire family. The account has less plausibility for the later cases, ones that were litigated even after the consensus was broadly accepted in the U.S., both socially and legally.

A second possibility is that the treating physicians actually disagreed with the family and thought that life prolonging interventions must be provided so long as there was any chance that the patient might survive, regardless of the quality of life. On this account, physicians saw death as an enemy that must always be fought as long as technology provided any means to fight it. On this account, the need for the consensus was due to a widespread physician misunderstanding of the role of medicine in prolonging life. Physicians did not realize, in the ways that patients and their surrogates did realize, that death is sometimes to be accepted as preferable to the agonies of a continued struggle to survive. Statement (1) of the consensus affirmed the validity of this patient-based understanding of death, and the rest of the consensus provided a series of mechanisms for carrying out the consensus.

The validity of this second explanation is supported by a study that my colleagues and I conducted in 1986 at the Houston Veteran's Agency Medical Center (Wray, 1988). We looked at the level of treatment provided to patients with very severe dementia when they were hospitalized, usually because of serious infections. In the majority of cases, the treating physicians treated the infections without discussing the issue with the families of the patients, even though they were often uncomfortable with providing the life prolonging interventions in light of the poor quality of life of the patients. The overwhelming reason for treating the infections was the fact that they were treatable, that the acute condition was reversible. I submit that this was a common attitude of physicians, reflecting an attitude towards technology and towards death that was common. The consensus was developed as a way of opposing this attitude.

The emergence of the appeal to futility grows out of a very different context. In this context, the physicians accept the fact that the patient is going to die and want to limit the use of interventions which at most prolong the dying process. It is the patient or the family that are demanding the use of the intervention.

This suggests to me the following possibility: as a result of the development of the consensus and of the extensive resulting discussions of the ethics of limiting life prolonging interventions, American (and other Western) physicians no longer see death as an enemy to be fought to the bitter end. They accept limiting life prolonging interventions at the request of patients or their surrogates, and they also want to be relieved of the burdens of providing those interventions when they won't work even if the patients or the surrogates want them to be provided. Many patients and their surrogates do not accept these views, and it is they who now see death as the enemy to be fought as long as technology provides means to continue the fight.

A recent study by Danis and colleagues (Danis, 1994) shed some light on the extent of these attitudes. Patients were asked twice (at an initial interview and two years later) about which interventions they would want if they had a terminal illness from which they would die no matter what the doctor did. Initially, over half still wanted such interventions as CPR, ICU care, artificial respiration, and tube feeding. Two years later, there was a decline in interest in these interventions in 42% of the participants but an increase in interest in 29% of the participants. As in other studies, those who had become more vulnerable (e.g., had been hospitalized, became more immobile) tended to want more of these interventions if they had a terminal illness. Many of the participants shared the attitude that death should be fought vigorously, even when they have a terminal illness. Of course, this study did not probe the depth of the commitment. Would they want the interventions if they only prolonged life for a very short period of time? Would they want the interventions if they only prolonged life in a comatose state? Nevertheless, it does suggest that the debate about futility is rooted in a correct perception that there are many patients and their surrogates who see death as an enemy to be fought as long as technology provides any means for fighting.

The consensus developed as a way of opposing a physician commitment to fighting death to the very end. Similarly, the debate about futility has developed as some seek a way of opposing a patient/surrogate commitment to fighting death to the very end.

As indicated above, Halevy and I have argued that it is considerations of integrity that should lead us to opposing that patient commitment. But that addresses this issue solely as a policy question, in the way that the development of the consensus treated physician commitment as a policy issue that could be resolved by the development of policies emphasizing patient autonomy. In the final section of this paper, I shall argue that these policies need to be supplemented by a much more extensive discussion of the very meaning of life and death. These discussions may lead to a reexamination of the attitudes that led in the past to physicians insisting, and in the present of patient/surrogates insisting, that life prolonging interventions must be provided as long as the patient continues to live.

III. THE MEANING OF LIFE AND DEATH

Since ancient times, Westerners have been divided into those who accept some form of belief in the afterlife, for whom the bodily death of the person is merely one more event in the ongoing history of the person who survives that death, and those who deny any existence of the person after his or her bodily death, for whom the bodily death of the person is the very end of their existence. I had in the past thought that this difference should also mark an important difference between how such people view both the meaning of life and death and the appropriateness of the continued use of life prolonging interventions when the prospects of their use are not very encouraging. My suspicion was that the former group, seeing bodily death as a transition to a new form of life (often viewed as a more blessed form of life), would reject in many cases the continued use of life prolonging interventions, while the latter group, seeing death as the very end of existence and the very end of all that is good, would insist on the continued use of life prolonging interventions as long as they held out some hope of success. This view turned out to be a mistake, both clinically and philosophically. Clinically, I have observed, in my practice as a clinical ethicist, no particular correlation between the degree of belief in an afterlife and the degree of willingness to accept the fact that one is dying and to reject further life prolonging interventions. Philosophically, I realized that decisions about life prolonging interventions might be based upon other beliefs. One might, for example, demand them in nearly all circumstances because of a belief in the infinite value of all human life, even while firmly believing in a blessed life with God after death. One might also forego them in many cases, based upon one's view that the resulting life might be worse than death, even while firmly believing that death is a tragedy because it marks the total end of one's existence. So it is not beliefs about the afterlife that I think that we need to examine.

What I want to propose instead is that there is a need to examine, as independently as possible of one's views about the afterlife, one's views about why life is a good and why death is something bad. In an important recent book, Frances Kamm (Kamm, 1993) has argued that there are a number of different components of those views about the evil of death. One is the deprivation factor, the fact that death robs us of the possibility of further goods. The second is the insult factor, the fact that death takes from us what we have, and thereby constitutes both a decline and an affirmation of our vulnerability. The third is the extinction factor, the fact that death means that our life is over. Obviously, the precise formulation of each of these factors will vary according to one's beliefs about an afterlife, but nevertheless all of these factors can be present in some form in our thinking regardless of our beliefs on the question of an afterlife.

Kamm goes on to suggest that all three of these factors might be modified by one's reflections on the goods of life. It is this suggestion that I want to explore, although my exploration of it will be somewhat different than hers. Like her, however, I want to emphasize in an account of the good life the importance of achievement goods (e.g., the personal relations with others one has had, the differences one has made in the life of one's community, one's moral and intellectual achievements) as opposed to experience goods (e.g., pleasurable experiences, experiences of satisfaction and contentment).

Some dying patients (particularly those in pain or requiring painful invasive interventions) are already willing to forego life prolonging measures because they see possible future goods as insufficient to justify the burdens of the continued interventions and of their continued existence. This type of reflection addresses, however, only the deprivation factor, leaving out the factors of insult and of extinction. Still others might be willing to forego further life prolonging interventions if they can be brought to recognize that a proper appreciation of their achievements and the extent to which they endure even after the person dies makes the extinction factor less important and less of a reason to insist on continued life prolonging interventions. Similarly, a recognition that their achievements are not diminished by their death, even if they are, makes the insult factor less of a reason to demand continued life through further technological interventions. Finally, a recognition that having more time to live is irrelevant to the goods that are really important in their lives, the achievements in which they take pride, unless that time is needed to complete those achievements or to attain others of a type not yet achieved, may further lessen the significance of the deprivation factor.

None of these reflections are meant to deny that death is often a tragedy, particularly when life ends before serious achievements have been accomplished, and nearly always involves some loss, since there are nearly always experience goods that could be, but will not be, attained. They are only meant to serve as the basis for a proper appreciation by those in whose life many achievements of a diverse kind have been accomplished that their death is not as great of a loss as it is sometimes seen as. Their achievements are neither diminished nor extinguished by their death, and death has not really deprived them of the opportunity to attain more of those goods that are most important in their lives. With such reflections, these people may be led to forego life prolonging interventions when the prospects of their success are small, because their death is much less of a loss than normally perceived. Moreover, seeing the patient's equanimity about his or her decision, the treating physician may be led to be at peace with limiting life prolonging interventions.

It seems to me that we need to encourage people to engage in these forms of reflection. People need to think both about what is really important in their life and what they have accomplished in regard to those things that are really important. It is most important for them to think about these issues from time to time when they clearly have time to live, even if only with the help of serious medical interventions, for doing so will help them focus on what is both important and as yet missing in their life. But it is also very important for people to think about these issues when they are facing death and life prolonging interventions offer little hope. Doing so may help people conclude that they really have no reason to request the provision of these interventions.

None of these reflections will eliminate the need for a social policy on futility, preferably one based upon considerations of integrity. There will be many for whom these reflections are irrelevant, either because they emphasize their life experiences rather than achievements or because their life is poor in achievements. There are those to whom these reflections will be relevant but who will not reflect upon them or be moved by them. All of these groups may be led in many cases to demand life prolonging interventions even when the likelihood of success is small because they see their death as a total loss and we will need a social policy to deal with those demands. But for those whose life has gone reasonably well in terms of achievements, these reflections may lead them to see their death in a new way, may enable them to accept their death with far greater equanimity, and may lead them to eschew the demand for futile life prolonging interventions.

In short, then, my suggestion is that enlightened reflection on the meaning of life and death may be of great value to individuals in making decisions about life prolonging interventions and may lessen, although certainly not eliminate, the need for a social policy on futility.

Center for Medical Ethics and Health Policy/Department of Philosophy
Baylor College of Medicine/Rice University
Houston, Texas U.S.A.

BIBLIOGRAPHY

American College of Physicians: 1992, 'Ethics manual: third edition', *Annals of Internal Medicine* **117**, 947–960.
Angell, M.: 1991, 'The case of Helga Wanglie', *New England Journal of Medicine* **325**, 511–512.
Blackhall, L.: 1987, 'Must we always use CPR', *New England Journal of Medicine* **317**, 1281–1284.
Brody, B. and Halevy, A.: 1995, 'Is futility a futile concept', *The Journal of Medicine and Philosophy* **20**(2), 123–144.
Council on Ethical and Judicial Affairs: 1991, 'Guidelines for the appropriate use of do-not-resuscitate orders', *JAMA* **265**, 1868–1871.
Danis, M. *et al.*: 1994, 'Stability of choices about life-sustaining treatments', *Annals of Internal Medicine* **120**, 567–573.
Halevy, A., and Brody, B.: 1996, 'A multi-institution collaborative policy on medical futility', *JAMA*, 276.
Kamm, F.: 1993, *Morality, Mortality Vol. 1*, Oxford University Press, New York.
Wray, N. *et al.*: 1988, 'Withholding medical treatment from severely demented patients', *Archives of Internal Medicine* **148**, 1980–1984.
Youngner, S.: 1988, 'Who defines futility?' *JAMA* **260**, 2094–2095.

SHIGEAKI HINOHARA

FACING DEATH THE JAPANESE WAY – CUSTOMS AND ETHOS

Compared with Western countries, the Japanese are a homogeneous group of people. However, this homogeneous group holds concepts of death that are extremely heterogeneous. This heterogeneity originates from the adoption of a variety of foreign cultures into the indigenous Japanese background. It is the end result of a process that occurred over a period of about two thousand years. Contemporary Japanese people conceptualize death in the following four categories (Yamamoto, 1988): (1) native culture, which originated from Shintoism; (2) Chinese culture, with Confucianism and Taoism as the main areas of thought; (3) Indian culture, as perceived through Buddhism; and (4) Western culture, with Christianity as central. These four ways of thinking exist in a mixture in the mind of the contemporary Japanese people, constituting their thoughts about death.

I. WEI ZHI

I will begin with a discussion of native Japanese thought on death. Its roots can be traced back to the period before the transmission of the continental Chinese culture.

Wei zhi (the Chronicle of the Wei Kingdom), written in the late third century, has a record of the Japanese people and there is an entry that describes the indigenous Japanese funeral. Through this record we are able to perceive the Japanese people's way of thinking about death. The record, translated by Tsunoda and Goodrich, indicates as follows.

When a person dies, they prepare a single coffin, without an outer one. They cover the graves with sand to make a mound. When death occurs, mourning is observed for more than ten days, during which period they do not eat meat. The head mourners wail and lament, while friends sing, dance, and drink liquor. When the funeral is over,

Kazumasa Hoshino (ed.), Japanese and Western Bioethics, 145–154.

all members of the whole family go into the water to cleanse themselves in a bath of purification (1951).

This idea that death is taboo, and something unclean, has persisted even unto the nineteenth and twentieth centuries in Japan. I was a child during the prewar time, when there were strict customs regarding death in the family. At the gate or the entrance to the house, a piece of rhombic shaped white paper was pasted. Inside this piece of paper was brush painted a character that means "taboo." The purpose of this custom was to display openly to the passers-by that there was a death in the house.

The body of the dead was regarded as unclean because once life is gone the body decomposes. Therefore, anyone who enters into the house also becomes unclean, as the unclean body is resting there. This belief brought about the custom of sprinkling salt, which purifies and exorcises, onto the visitor when he leaves the house of the dead. Now a small pouch of salt is included along with a gift of thanks to those who come to a funeral. Thus the formality of purifying death with salt has persisted to the present.

II. FOUR VIEWS OF DEATH

During the lifetime of Prince Shōtoku (574–622), the Japanese adopted Taoism from China, along with Confucianism. Taoism is a teaching propagated by Lao tzu and Chuang tzu. In Taoist thought a human being is a part of natural phenomena. Therefore, the best way for a human being to live is to follow the course of nature. The representative Japanese who took up Taoistic thought and lived it is Ryokan, a Zen Monk, who lived from 1758 to 1831.

Confucianism, on the other hand, is a teaching of Confucius of China (552–479 BC). This thought holds rites of loyalty and filial piety especially high. Bushidō, the Way of the Samurai Warrior of Japan, was cultivated under the influence of Confucianism. In the Japanese concept of Bushidō, the supreme death is suicide, Junshi, self-determination. Two styles of suicide have been regarded most highly by the Japanese through out history: Seppuku, suicide by disembowelment, and Junshi, self-immolation to follow one's master in death.

Junshi is considered to be an ethical action. It is a dedication of the entirety of one's life to an admired person. This kind of suicide was accompanied by a poem. This action was referred to as a "leaving life – jisei" expression, and it portrayed the feelings of one who either accepted death, or transcended death. These last words have also been valued in Japan, as a way of indicating the dying person's philosophy of life, or his ethical perspective.

The third view of death held by the contemporary Japanese is influenced by Buddhism, which developed from the Indian Hindu religion. The Japanese inherited a Buddhist idea of transmigration and rebirth. There are "nenbutsu" sects that believe in chanting. The believers call out the name of the Amida Buddha. The sects are widely known also as Pure Land Sect, True Pure Land Sect, or "Ji" (Timely) sect. The believers of these sects wish for rebirth after death in the pure land, or paradise, by chanting "namu amida butsu – save us, merciful Buddha." these sects were popularized by Buddhist high monks such as Honen (1133–1212), Shinran (1173–1262), and Ippen (1239–1289). Through these Buddhist leaders, rebirth in paradise by chanting became possible for the lay public.

The Zen sect, also Buddhist, taught somewhat differently. A believer attains enlightenment which transcends life and death through meditation. A person in pursuit of the Way trains himself in a Zen temple. He obtains at the end a perspective on life and thanatology, a perspective on death, which altogether transcends worldly ideas on life and death.

The fourth view of death held by contemporary Japanese people has been influenced by Western thought. Christianity was transmitted to Japan during the Muromachi period, 447 years ago (1549), by Francis Xavier. This teaching conveys the concepts of original sin, and believes that there is a last judgment after death. Salvation and eternal life are attained through living life in awe and reverence of God, about whom one learns through the teachings of Jesus Christ.

III. THE JAPANESE ETHOS

Modern Japanese people cultivate a philosophy of life and thanatology that fits into one of the above, if not all of the above, categories. Even so, there are some native ideas that have persisted throughout history and are common to all of the Japanese people. For example, resignation to death rather than fighting it, is one such idea. Mind and body are in being one, which means that they are inseparable, expressed as "shinshin ichinyo." A life style that values tasteful simplicity (wabi), or austerity (sabi), which is expressed in the tea ceremony, is an example. Another example can be found in the mind set that pursues experience and learning. Here, one believes in "ichigo ichie," that each meeting, or experience, is the first and last in life.

What should be mentioned and emphasized is a thanatology that permeates the essays of Kenkō Yoshida , called *Tsurezure-gusa, Essays in Idleness*. The essays were written in the fourteenth century, and present images of death that still influence the contemporary Japanese people. The essay, translated by Donald Keene, comments as follows.

> The four seasons, after all, have an appointed order. The hour of death waits not its turn. Death does not necessarily come from the front; it may be stealthily planning an attack from behind. Everyone knows of death, but it comes unexpectedly, when people feel they still have time, that death is not imminent. It is like the dry flats that stretch far out into the sea, only for the tide suddenly to flood over them onto the shore (Keen, 1967, p. 138).

These are indeed superb words on human death, which creeps up behind us. Kenko shows powerfully that the ordinary Japanese do not realize this fact. Most contemporary Japanese people have a tendency to forget Kenko's wisdom in their own everyday life, as long as they and their close kin are healthy. Only once they experience a death in the family do they focus their thought on their own forthcoming death. At such times death gains reality in their eyes.

The novelist Yasushi Inoue (1907–1991), wrote an essay at the age of 58 titled: "My late father." It states:

> When my father was well, I never once thought of my own death. Since even my father was alive, I was, so I felt, far from my own death. Only after I lost him did I realize, for the first time, that I had been escaping from death behind a thin screen which guarded me had been removed. Having lost him, I saw before me, for the first time, the surface of the sea of death. Suddenly there was no barrier between it and myself, and I felt curiously desolate (Inove).

For the Japanese in general, it is usually the case that their perspective on death changes with age. A young person knows intellectually that he or she is walking every day towards the ultimate goal of death. But most of them do not think consciously that death is coming near.

The elderly person's feelings about death is depicted by the writer Hori Hidehiko (1902–1982), in "At the bank of the river death."

> Till I turned 70, I thought it was I who was nearing close to death year by year. Therefore to keep on living or to die was a matter of my choice, so it seemed to me. I thought I could commit suicide any time I liked. Like Hamlet, I thought living and dying were within my power.

> Now that I am 82 years old, I feel death is coming towards me from the front, step by step, without taking me into account. It is not I who am nearing death day by day. Death is closing in on me.
>
> This was an unexpected change of perspective for me. To put it another way, it is not that I am going to die. Death is coming to me one day, and will kill me (Hori, 1987).

Such fear towards death definitely exists among the Japanese elderly. However, the words and actions of elder artists and poets of Japan show the conceptual intermingling of life and death in a most beautiful way. They have an outlook on life that is objective, which enables them to live their lives to the fullest. There have been many such people.

Saigyō, for example, a wandering monk and poet (1118–1190), lived in such a manner. Saigyō's lifestyle is known to many of the Japanese who are over 70, who lived before and during the second World War. Here follows one of his poems.

> If I had my wish, I would die in the spring,
> beneath the cherry blossoms,
> under the full moon.

Saigyō left the secular world and became a monk at the age of 23. His wish was to die in the spring month of February by the lunar calendar, at the time of the full moon. He died at the age of 72.

Dr. Shigeji Kijima, a close friend and classmate of mine at the Kyoto University Medical School, died in 1984 at the age of 71. Three years before his death, he took his own biopsy specimen from the rectum. He examined the specimen under the microscope, promptly diagnosed cancer of the rectum, and underwent an operation. Three years later the cancer metastasized, and he died. A week before his death, critically ill, he dictated a ten page message from his bed. It was titled "Terminal care at the hospital." Here is an excerpt:

> . . . to care for patients who are suffering from anxiety, one must consider methods of nursing. First, we should anticipate oncoming pain. If the patient is afraid of pain, we must explain how modern medicine can remove it. We should try to relieve the patient of unnecessary worry . . . we should design terminal care in such a way so that any patient who can work or have hobbies, may carry that out as best as he can. This is not only a distraction from the suffering. It will make the patient experience the fullness of life by creating something meaningful . . . (Kijima, 1984).

In addition to Dr. Kijima's suggestions, if no medical cure is available, one way to help the patient accept death peacefully is to invite a clergy person whom the patient trusts. This is if the patient has an interest in religion, and wishes to speak to this type of person to receive spiritual care. In Western countries, it is commonplace for terminal patients to receive the last sacraments and so on through hospital chaplains, priests, ministers, rabbis and other clergy. But in Japan, as a custom, hospitals as well as patients do not accept visits by Buddhist monks.

Dr. Kijima, when death was near, called his three children to his bedside. To his first son he taught that "sincerity/heart" was important. He advised his second son to value his home life more. To his daughter he promised "gue issho," to meet again after death, in paradise. His faith is expressed by the term "gue issho," which in Buddhist terms means that Amida Buddha is reborn in the Pure Land, and welcomes all the believers who are reborn there. Because of this Buddhist faith, he was able to live determinedly in the face of death.

I visited Dr. Kijima a few days before he passed away, at the hospital in Hamamatsu city of which he was president. When I opened the door to take leave, he raised a hand, and said: "I will go a bit earlier than you."

Zen is another sect of Buddhism. But unlike the Pure Land sect, it preaches that life is emptiness, and life and death are intertwined on the same level. It takes a monistic view and believes that mind and body are of one origin. Professor Daisetsu Suzuki was a highly acclaimed Zen scholar both in Japan and overseas, for his dissemination of Zen thought. Professor Suzuki was taken to St. Lukes's International Hospital by ambulance for an obstruction of the intestine, called ileus. When he came into the hospital his blood pressure declined due to shock, which made emergency surgery impossible. Only pain treatment, using morphine, was given as palliative care. Two hours before his death, I asked him:

"Would you like to meet the leaders of the temple?"

Professor Suzuki answered:

"No, I don't have to meet anyone. I am alright, alone."

Professor Suzuki's secretary, who called him "Sensei-Master," sat by his bedside. Later she told the disciples of Professor Suzuki: "Professor Suzuki lay there motionless, but it seemed to me a continuation of his living. I did not feel that a great change took place between Master as he was alive, and Master as he passed away."

He valued every moment, and did his best for here and now. For him, even death was "nothing special." Such Zen thought applies also to the Japanese tea ceremony and Haiku poetry composition. More recently a haiku poetess Takajo Mitsuhashi (1899–1972), mused as follows.

White glistening dew;
on the day I die,
I shall tie an obi-sash (Mitsuhashi, 1980).

This haiku poem describes objectively and lightly the calmness in the everyday life of this poetess, who was sick in bed. She lived whole, constantly communicating with her own death.

Our deaths vary, depending on the individual. In other words, one's death is completely individual, as unique as the person himself. A physician has the tendency to count a patient's death as one of many that occur among his patients. But that should not be the case; he should not regard a patient as a case of disease, only paying attention to his condition as a scientific object of study. Rather, a physician must value the personality of the patient, as a person who is facing death and pursuing his own life in a unique manner. As long as the patient is alive, the physician should make the maximum effort to give the patient a better quality of life. It is said that medicine is not a science, but an art. This art refers to the role that a physician must play in caring for the patient so that he may accept his individual death, in his own way.

IV. DEATH AND THE GRAVEYARD

Before the transmission of foreign cultures, our ancestors believed that the spirit of the dead persisted in some form in the world. In this way of thinking, a human being is not terminated by physical dead. Thus came the mentality of revering the graveyard, where the dead rest. The Japanese called July "the month of those who passed away," and the thirteenth to the fifteenth of this month is called "Bon." At this time all family members customarily visit the graveyard.

At the time of "Bon" one lights a fire to welcome the ancestral spirit. The ancestral spirit is also sent off with a fire, an example of which is "toro nagashi," a floating paper lantern, lighted by a candle, which floats down a river. This is done to see off the spirit. Yet there was no such

"Bon" festival custom in Buddhism; it was the native Japanese custom of ancestor worship that was adopted into various sects of Buddhism and Shintoism. In a common non-Christian Japanese household not only flowers but also food is offered at the Buddhist family altar. Also, it is common even for Christians to visit the family graveyard at the time of "Bon."

It has been said that the majority of the contemporary Japanese people are not religious. But the custom of being aware of the ancestral spirit is established firmly in the national lifestyle, through various events of ancestor worship and graveyard visit. In this sense, the Japanese hold steadfastly to the ethical custom of valuing life. This is seen through the rites of funerals, as well as other occasions of remembrances of the dead. About 80% of funerals are Buddhist, 10% are Shinto and 5% Christian.

The *Yomiuri Newspaper* carried out a nation-wide opinion poll in 1989, regarding people's religions. To the question "Do you believe in a religion?" 26% responded "yes," and 72% "no." In contrast, according to 1992 statistics, the total number of believers registered by religious organizations of Japan adds up to more than 220 million, two times the actual population.

A farewell message is read aloud at the Japanese funeral. This has the important function of impressing onto the mourners that the spirit of the dead is present. Since a farewell message calls out to the spirit, it is read facing the body or the ashes. Subsequently incense is burnt and flowers are offered by the people. In many cases, Christians perform without hesitation this Buddhist custom of burning incense. In other words, Japanese individuals attend Buddhist funerals without giving any thought to personal religious beliefs. The Japanese people are accustomed to attend funeral formalities of all religions. They do this regardless of their own religious convictions, because they consider it only courteous toward the dead and his family.

Many Japanese worship the spirit of the dead, even if he or she has no religious faith. This they do through attending various death rites. By so doing, he or she cannot ignore the spirit of the dead. What makes them different from Western people, in this context, is that the Japanese are more individualistic and do not accept as readily the religious rites of others. The reason for this attitude of the Japanese is their high regard for the ancient native lifestyles. They are viewed as important indigenous customs and events related to death. This attitude has been cultivated primarily at home, and then in the community. Within the context of these customs, the individual and collective ethics of Japan have been constructed.

V. CONCLUSION

Professor Tetsurō Watsuji (1889–1960), founded the Japanese Association of Ethics in 1950. He defined the Japanese term ethics, "rinri" as follows. "Rin" refers to the relationship between human beings, and "ri" as the way of living together that evolved among human beings. Thus "rinri" is ethics.

The Japanese people's ethics were closely related to the customary events held in the community. Ethics, the way of life of the Japanese people, is influenced strongly by thanatology based on personal death, Buddhist funeral, and subsequent memorial rites. Thanatology has developed from the multiple factors stated above. In recent times, these views are being strongly challenged and influenced by Western cultures. In the near future, the thanatology of the Japanese will change rapidly due to the following factors. (1) There is a tendency towards obtaining informed consent on the part of both the physician and the patient. (2) Patients are becoming more aware that they can act and make decisions independently. Presently, the ethical act of telling the truth to patients with terminal cancer is secondary to the wishes of the family. The majority of Japanese physicians also resist telling the truth to advanced cancer patients, regarding the act as snatching away from them any hope for life. This resistance to telling the patient about terminal cancer seems more prevalent among young physicians who advocate a scientific approach. (3) A change is taking place regarding funeral customs. Attention must be paid to the current resistance against tradition that is developing among the intellectuals, writers and artists of Japan. These people have not been religious throughout their lives. Therefore a customary funeral, or a funeral held in a religion whose faith they do not share is repulsive to them. The more famous of such people announce: "When I die, no wake, funeral, or farewell ceremony should be held."

At times this attitude reflects an expression of the Japanese traditional preference for a frugal and sincere style of life and human relationships. At other times it reflects a repulsion towards superficial funeral rites. Recently memorial ceremonies have become more important. A major challenge remains, though, in attempting to incorporate and reflect

Japanese traditional customs, as well as ethics regarding the dead in a concrete manner.

St. Luke's College of Nursing
Tokyo, Japan

BIBLIOGRAPHY

Hori, H.: 1987, *Shi no kawa no hotori nite*, Sakuhinsha, Tokyo.

Inove, Y.: 1983, 'Unforgettable people (Wasurerarenu hitobito)', *Inoue Yasushi Essay Collection*, Gakushu-Kenkyusha, p. 1.

Keen, D. (trans.): 1967, *Essays in Idleness*, Columbia University Press, New York.

Kijima, S.: 1984, 'The terminal care in a hospital', *Shoken (The Academic Annual Report of Shizuoka Rosai Hospital).*

Mitsuhashi, T.: 1980, *The Collection of Modern Women's Haiku Poems*, Kodansha, Tokyo, p. 2.

The Yamiuri Newspaper, October 11, 1989.

Tsunoda, R. and Goodrich L.C. (trans.): 1951, *Wei zhi (The Chronicle of the Wei Kingdom)*, P.D. and Ione Perkins, South Pasadena.

Yamamoto, S.: 1988, *Nihonjin no Shiseikan Shiseigaku (Thanatology)*, Gijutsu Shuppan, Tokyo.

GEN OHI

ETHOS AND ITS CHANGES: A COMMENTARY ON 'FACING DEATH THE JAPANESE WAY – CUSTOMS AND ETHOS'

Dr. Hinohara analyzes the history of the Japanese view of death. His essay speaks well of his excellent insight as a historian into Japanese culture. In his conclusion, he predicted that the Japanese ethos regarding ethical sentiments about death would undergo rapid changes. I propose to focus on ethical sentiment regarding information provided by a doctor to a patient of a disease who has a poor prognosis for recovery (e.g., an incurable cancer.)

Life in developing countries differs from place to place, but a common feature is the strategy for survival with mutual support provided from within a group to which a person belongs (such as a family, tribe, or community.) Life in rural Java in Indonesia typically reminds us of Japanese village life in historical times. If a person makes a cake, she or he must make enough for all of her/his significant others. Everyone is obliged to assist those in need of help in farming the fields. In return, one expects help from others when in need. Everyone knows what is going on in everyone else's life. It is easily assumed that an awareness of a concept of privacy is rarely formed in such a society. A Japanese student once complained to me that her fellow students from Indonesia never knock when they enter her room. She did not realize that it is only in recent years that the Japanese began knocking on the door prior to entering a room. At one time, there was no Japanese equivalent for the word "privacy."

Interpretations of "person" and "ethos" observed in a society that has long depended on mutual support are quite different from those which develop from within a society in which individualistic ways of life are considered normal. As for the interpretation of "person," it is well known that a distinction of the self from others or a definition of

Kazumasa Hoshino (ed.), Japanese and Western Bioethics, 155–159.

an individual is more obscured in traditional Japanese society, rather than in a society that considers autonomous living to be natural. One's self is formed and owned in part by significant others. Such a state may be perceived as unbearable interference by those living in a modern society where self and others are clearly distinguished, or by those who consider it natural that distinctly defined individuals should be armed with concepts such as "rights," "self-determination," and "privacy." At the same time, such an individual is unaware of his loss of the deep-rooted empathy that underlies thoughtfulness toward others. Such empathy plays an essential role in a mutually supportive society. To those accustomed to a mutually supportive life, the individualistic life of Western society is very cold.

As linguists point out, intercultural communication pertaining to the domain of perception and emotion is basically unattainable, while common understanding can be achieved over anything that is logically analyzable. It appears that this dichotomy lies at the base of the difficulty in translating the concept of "person" across heterogeneous cultures.

It also seems that the ethos of a mutually supportive society tends to move in the direction of non-maleficence and beneficence, as these concepts are defined by ethicists. In other words, a mutually supportive society values the maintaining of good human relations. If a physician's attitude regarding truth disclosure concerning incurable diseases is to reflect the ethos that provides the underpinning of the society, one should expect the doctor within a traditional mutually supportive society to favor non-maleficence and not to disclose the dismal truth about the illness to the patient. In many countries of the East and Southeast Asia, such as China, Taiwan, Cambodia, Thailand, Malaysia and Indonesia, patients are not always informed of incurable cancer. This is true of all typical physicians irrespective of religion. The ethos regarding information about fatal disease in countries south of the Sahara Desert also seems to resemble that of Asia. An American physician visiting Uganda, for instance, where one fifth of the population of reproductive age is infected with HIV, would be extremely surprised to hear that these patients are not typically informed that they have AIDS.

What about Europe? According to a questionnaire survey of 260 gastroenterologists in 28 countries, reported in *The Lancet* in 1994, physicians in Germany and Holland responded that they would tell the truth to patients with colon cancer, however, only about half of those surveyed from the United Kingdom, France, and Italy indicated they would tell the truth. Those in South, East and Central Europe, including Spain and Russia, indicated that they would not inform the patient. The countries where a physician would inform his/her patient are commonly characterized by Protestantism and affluence during the past several centuries. It is possible to interpret these countries has having had the socioeconomic freedom to lead the relatively autonomous lives to which many hold fast.

Japanese ethos appears to be moving toward respecting autonomy, as least in urban areas. How fast is this change occurring in the country as a whole? Most information regarding medical ethics in Japan has been imported from the United States. According to Novack *et al.*, only 10% of American physicians supported informing patients of cancer in 1961. However, by 1977, 97% supported informing patients. According to the surveys on physicians' attitudes regarding truth disclosure to cancer patients, conducted by the Asahi newspaper in Japan in 1980 and 1989, the percentages of those physicians who agreed with informing patients of their incurable cancer remained unchanged at 37%. Contra the United States, there was no change despite the ten year time lapse. In practice very few physicians actually impart the truth to such patients.

Another factor which reflects the Japanese ethos regarding terminal care is admission to a hospice. In the United States there were about 1,400 hospices in 1984, ten years after the first hospice was introduced in 1974. In Japan, however, only 20 hospices were operating a decade after the first was established in 1980. This suggests that while American phenomena such as jeans, rock music and baseball were enthusiastically and rapidly accepted in Japan, change to the ethos of truth disclosure has been very gradual, to say the least.

What would be the speed of future changes to this ethos? There is no adequate date to predict with. However, I would like to raise the following three points. First, truth telling assumes autonomy on the part of the patient. It assumes that he, himself, is able to make the best possible choice for himself, even if the information is not welcome. Most cancer patients are elderly. According to our survey in Okinawa, the ability to make decisions lowers in elderly people with impaired intelligence. For instance, our survey in Nago City revealed impaired intelligence in 144 of 223 elderly persons studied, even though none were obviously demented or had any trouble making decisions in daily life. When asked to choose between "life prolongation" and "pain removal" in terminal care, 78% of the normal group desired to have their pains removed.

This coincided with the result of the middle aged group also surveyed. However, in the group with impaired intelligence, the ratio changed to 34%. While people with lowered intelligence may want to live longer, it is more natural to interpret the data as reflecting that they are no longer able to make adequate choices. It is difficult to detect an intellectual impairment in the elderly so long as it is clinically compensated. An imposed knowledge of having incurable cancer would certainly be a burden that they are much less likely to be able to handle. Since the prevalence of intellectual impairment rapidly grows with aging, it is morally questionable to inform such persons with clinically compensated intellectual impairment and restricted autonomy that they have incurable cancer.

Second, a person begins to develop ethical awareness in infancy. When a Japanese child starts school he/she is taught by his/her mother to "get along with others." This is clearly an ethical orientation, at the core of which lies the will to coexist with others as expressed in terms of non-maleficence and beneficence. In other words, it means that one undertake efforts not to harm others and to maintain harmony with the whole by restraining oneself. It is interesting to note that in the United States, a child is inculcated by his parents to be autonomous, to "judge and determine for himself." Those familiar with the ethoses of both Japan and the United States would hardly question the point that these parental admonitions each represent an optimal strategy to ensure the child's probability of success in the respective society. At the same time, they will morally condone the Japanese physician who has been given the ethical education emphasizing non-maleficence in their childhood, and who show stronger resistance toward giving information about incurable disease to patients than their American counterparts.

Finally, I would like to discuss the issue of trust. One strategy to improve human relationships is empirically based on a behavioral rule that not harming others and doing good for them enhances the probability of personal survival. Underlying this rule is trust in others. Trust involves allowing others to take care of items the individual cares about. It is quite probable that elderly persons who have lived in a society dominated by the ethical sentiment of mutual support tend to consign to others more than those who have lived in a society where they make all their decisions on their own.

In summation, the survival strategies that have been adopted in the United States and Japan, the dominant ethics in each culture, are oriented in different directions. As Japanese society is increasingly exposed to heterogeneous cultures, it is beginning to accept cultural diversity with unprecedented speed. Concerted research efforts in the fields of medicine, anthropology and philosophy are warranted to elucidate how Japan's ethos will change in the future.

School of International Health
The University of Tokyo
Tokyo, Japan

BIBLIOGRAPHY

Hinohara, S.: 1996, 'Facing death the Japanese way – Customs and ethos', in this volume, pp. 145–154.

AKIRA AKABAYASHI

THE CONCEPT OF HAPPINESS IN ORIENTAL THOUGHT AND ITS SIGNIFICANCE IN CLINICAL MEDICINE

I. INTRODUCTION

"Quality of life" has become an important notion in clinical research. According to the international medical database "Mediline," the number of articles that have "quality of life" as a keyword have increased from 370 per year in 1983 to 1130 per year in 1993. The Japanese medical database "Japana Centra Revno Medicina" shows an even more dramatic increase: 440 articles per year in 1993 in contrast with 23 articles per year in 1987. This growing interest in the concept of quality of life involves a variety of medical fields, such as the treatment of cancer patients, care for psychiatric patients, care for the elderly, etc.

However, it seems that most clinical researchers have managed to avoid addressing a relationship between notions of quality of life and the concept of happiness. Happiness, as a major moral concept, is ethically and culturally at home in Japan and other Asian nations. Without a definitive understanding of happiness, we will not be able appropriately to deal with a variety of issues relating to quality of life.[1]

Accordingly, when the issues of quality of life are deliberated in Japan and other Asian nations, I would like to point out an agenda item of some importance. That is, we must begin by undertaking a close examination of the traditional concept of happiness. The moral spirituality of modern Asian people has certainly been influenced by the traditional manner of general Oriental thinking. This is the case in particular for understanding the concept of quality of life in reference to the traditional concept of happiness.

As is well-known, Asia is a place where Buddhism, Christianity, Islam, Confucianism, Hinduism, and Taoism have evolved and diverged. Besides these world religions, various folk-faiths also prevail. This type of coexistence of a variety of religions and faiths constitutes one prominent characteristic of Asian countries. Since Japanese people have

Kazumasa Hoshino (ed.), Japanese and Western Bioethics, 161–164.

been tremendously influenced by Buddhism, the rest of this paper begins with an examination of the Buddhist understanding of happiness. Then the question of the significance of happiness for medical professionals will be addressed.

II. THE BUDDHIST UNDERSTANDING OF HAPPINESS

Let me begin by asking what the relationship is between *nirvāṇa* (nehan) and happiness in Buddhist thought. *Nirvāṇa* is the final goal of Buddhist aspiration and practice. It frees one from suffering, death, rebirth, and all other worldly bonds. It is the highest transcendent consciousness. In the understanding of early Buddhism, *nirvāṇa* is a departure from the cycle of rebirths and entry into an entirely different mode of existence. It requires overcoming completely the three unwholesome roots – desire, hatred, and delusion – and the coming to rest of active volition. *Nirvāṇa* is not seen in a positive relation to the world but is only a place of salvation. For Buddhism, which sees all existence as ridden with suffering, *nirvāṇa* is interpreted as the cessation of suffering, and suffices as a goal for spiritual effort (Schuhmacher and Woerner, 1994; *Japanese–English Buddhist Dictionary*, 1991).

Yutaka Yuda summarized the interpretation of happiness in relation to *nirvāṇa* throughout the history of Buddhist thought:

> Buddha mentioned happiness in *Dhammapada*, one of the oldest collections of his moral teachings . . . "Hunger is the worst disease, the body itself is the highest suffering. If one accepts this, *nirvana* is the happiest state (*Dhammapada*, 203)." Buddha regarded health, a state free from diseases, as supremely good, and one of the highest achievements of happiness . . . In early Buddhism, *nirvāṇa* was thus regarded as eternal happiness (Yuda, 1983, pp. 73–89).

Later in *mahāyāna* (Great Vehicle), one of the great schools of Buddhism, *nirvāṇa* denotes non-production and non-destruction, and is equated with wisdom and the *dharma-kāya* (the true nature of Buddha). In this context, *nirvāṇa* has four essential qualities: eternity, happiness, substantiality, and pureness.

In *mahāyāna*, a *bodhisattva* (bosatsu) is a being who seeks buddhahood through the systematic practice of the perfect virtues, but renounces complete entry into *nirvāṇa* until all beings are saved (Yuda, 1983, pp. 151–159). The determining factors for his action are compassion and altruism, supported by highest insight and wisdom. The *bodhisattva* provide active help and are ready to take upon themselves the suffering of all other beings. Thus, if people ask for any help, they will not be refused it.

In *Tariki-nembutsu* (Pure Land Buddhism), *Amitābha* (Amida) Buddha grants one's entrance to the Pure Lands as soon as one has faith in him. All one has to do is *nembutsu*, a form of meditation upon *Amitābha*, in which one attempts to become fully aware of his mercy, and then totally surrender to *Amitābha*. In short, according to this faith, a transcendent existence would bring welfare or happiness to people.

III. THE SIGNIFICANCE OF HAPPINESS FOR MEDICAL PROFESSIONALS

One issue relating to the concept of happiness is not frequently discussed. That is the issue of the meaning and significance of happiness for care givers or medical professionals. This is an important topic for those who work in the day to day reality of the clinical setting. What is good for a patient and what happiness means to a patient are often discussed. One of the goals of clinical medicine which is repeatedly stressed is the provision of quality care to patients. However, there has been little discussion on what kind of conduct provides happiness to medical professionals and care givers. Raymond Devettere's essay entitled 'Clinical ethics and happiness' has important implications for this issue. He holds that the personal happiness of the clinician, rightly understood, is a legitimate and important goal of clinical ethics.

> More important, perhaps, than any attempt to answer the criticism that can be brought against eudaimonism in ethics is a simple truism. If living ethically does not achieve the greatest possible good for me, if the goal of ethics is not my personal happiness, if the ends I pursue in ethics are not agent-relative but agent-neutral, then ethics is something I can easily set aside ... In clinical ethics this means that it is not enough to say the goal is better patient care, what is good for the patient. The goal of clinical ethics centers on what is good for the clinician – personal happiness rightly understood (Devettere, 1993, p. 87).

This rightly suggests that the concept of happiness should have a broader usage in clinical setting. Considerations of happiness should extend to health care providers, not only to patients.

IV. POSTSCRIPT

A question remains. What does happiness mean to us in the present? If conceptions of happiness vary with eras, cultures, religions, etc., does this mean that we have to establish a concept or theory of happiness in our own period within our own cultural setting? Is there a universal concept of happiness for all human beings in all times? If there is one, then what is it? In any case, re-examining and understanding the traditional concept of happiness should be helpful for us to come to terms with the concept of "quality of life" in clinical research.

Department of Ethics and Anthropology
School of International Health
The University of Tokyo
Tokyo, Japan

NOTE

[1] Most clinical researchers try to avoid offering an exact definition of happiness. But this is the case not exclusively in the field of health care. For example, a book entitled *Does Science Make People Happy?* (Utumi, Ookuma and Kato, 1989) is written by 20 authors from a variety of disciplines. The editors have defined happiness broadly because they believe that the definition of happiness for a person depends upon that person's values.

BIBLIOGRAPHY

Devettere, R.J.: 1993, 'Clinical ethics and happiness', *The Journal of Medicine and Philosophy* **18**, 71–89.
Japanese–English Buddhist Dictionary: 1991, Daito-Shuppansha, Tokyo.
Schuhmacher, S. and Woerner, G. (eds.): 1994, *The Encyclopedia of Eastern Philosophy and Religion: Buddhism, Hinduism, Taoism, Zen*, Shambhala, Boston.
Utumi, J., Ookuma, Y. and Kato, I. (eds.): 1989, *Does Science Make People Happy*? Keisoshobo, Tokyo, pp. i–vii.
Yuda, Y.: 1983, *A History of Buddhism*, Hokujushuppan, Tokyo.

PART FOUR

THE HUMAN GENOME: TABOOS AND MORAL INTUITIONS

KURT BAYERTZ

THE NORMATIVE STATUS OF THE HUMAN GENOME: A EUROPEAN PERSPECTIVE[1]

I. INTRODUCTION

To the same extent as progress in modern gene technology, as well as the increasing number of its possible applications, which have made the human genome accessible to both diagnostic and manipulative intervention, the normative status of the human genome has become a directly practical, as well as a theoretical problem. For years, it and connected problems have been the subject of debate, not only amongst academics, but also politicians and lawyers. In this paper, I shall outline some of the main European tendencies within this debate. My goal is to show that, despite all of the controversies, something akin to a specifically European "moral culture" is shining through this debate. This moral culture differs not only (which should be no surprise) from Japanese approaches to bioethical problems, but also from main tendencies of bioethical thought in the U.S. This specifically European moral culture is partly to be found within the documents which have been written over the past two decades about bioethical problems on a European level. There have been major efforts by the *European Community* and by the *Council of Europe* to reach a consensus about bioethical issues and policies among their member countries. Especially important is the work of the Parliamentary Assembly of the *Council of Europe* and a group of experts for the preparation of a *Convention for the Protection of Human Rights and Dignity of the Human Being with Regard to the Application of Biology and Medicine: Bioethics Convention*, the preliminary draft of which was published in June, 1994. Work on this first international convention for bioethical matters has not yet been completed; the committee has, however, been able to reach a consensus in some important matters, and a compromise in others.

I will summarize some important results of these European discussions, insofar as they are relevant to the problem of the normative

Kazumasa Hoshino (ed.), Japanese and Western Bioethics, 167–180.

status of the human genome. Moreover, I shall pay special attention to Germany and other German-speaking countries. These countries are especially significant in their fundamental objections to some of the new biological options available. These objections have in some cases even become legislation. Germany and Switzerland, for example, have outlawed a group of biotechnological possibilities which are seen as incompatible with the constitutional values of these states. In Germany, the *Embryonenschutzgesetz* (Embryo Protection Bill), which came into power in 1990, prohibits (imprisonment of up to five years or a fine) in §5 any intervention in the human germ-line, in §6 the cloning of human beings, and in §7 the creation of human-animal hybrids. Switzerland has even included a ban of this kind within its Constitution, following a public vote on the 17th of May, 1992. The reasons given for these restrictions may be interpreted, at least in part, as the expression of culturally specific moral convictions.

Yet can a homogenous "moral culture" exist within modern pluralistic societies at all? Of course, German and German-speaking views about the normative status of the human genome (as about nearly every other moral question) are just as far from being homogenous as they are within the whole of Europe. Even if there are widespread (but not unanimous) reservations about any kind of intervention in the human germ line, the reasons given for these reservations are very different (Mauron and Thévoz, 1991). Talk of a "moral culture" therefore has to be misleading if aimed at more than the identification of several, particularly influential lines of discussion. In order to identify such lines, I shall limit myself within the following to a certain type of argument: all of those based primarily upon the law. Law, and in particular the norms and values underlying a Constitution, may be viewed as the "secular religion of modern societies," with their directions possibly having the greatest claim to general acceptance. They are also the norms and values most likely to influence legislation and therefore medical practice. There is actually a special type of reasoning situated between the professional legal debate, on the one hand, and the general ethical debate, on the other; within it, argumentation oscillates between (Constitutional) law and morality. Concentrating on this type of argumentation implies not only that certain positions and arguments (regardless of their philosophical significance) have to be ignored, but also that my portrayal of European moral culture will mainly be descriptive, in the sense that not every trait of this portrayal will coincide with my own opinion.

There are four basic Constitutional principles which may be possible candidates for ways of justifying the normative status of the human genome: (1) the right to ownership; (2) the right to freedom from injury; (3) the right to unfold freely one's personality; and (4) human dignity. First of all, I would like to take a look at the first two of these principles and attempt to justify why they are only partly instrumental to the problem of the normative status of the human genome. I will then turn to the other two principles and emphasize the particular role played by human dignity. Finally, I will give a short summary and draw some conclusions.

II. BODY AND OWNERSHIP

The human genome is part of the human body, and for this reason alone it has a normative status. The disenchantment and moral neutralization of external nature which have risen with the modern sciences have never completely taken hold over human nature. For this reason, the human body is not morally neutral in the same way as the rest of nature. It represents the biological basis of the human personality and subjectivity, and is, to this extent, closely connected with them. Interventions in the body without the agreement of the person concerned are therefore neither morally nor legally permissible. There can be no doubt that the human genome is part of the human body, and there can, therefore, be no doubt that it possesses the same normative status. The "right to freedom from bodily injury," for example, prohibits the removal of genetic material from a non-consenting human being. The only exceptions to this rule are to be found in the course of criminal trials. It is permissible for a judge to order that genetic material be removed against the will of the accused for the purpose of identifying a person as an offender. Using genetic material for diagnostic, therapeutic or other purposes without one's consent would violate the ownership rights of the person. The categories "body" and "ownership" therefore also guarantee the human genome a normative status.

There are two reasons, however, why the right to freedom from bodily injury and the right to ownership play only a minor role in the European debate on the normative status of the human genome. The first of these reasons is concerned with the biological specificity of the genome. Although, "ontologically" speaking, the genome is part of the body, biologically speaking it assumes a special position within it. The genome is not one particular "organ" but functions as "building instructions" for the entire organism, including all of its organs. From a biological point of view, a line must be drawn between the genome and the body. In other words, although genetic material is part of the body, genetic information is not. True, it is impossible to separate genetic information from genetic material, yet we have to examine the relation between the two. Does ownership of genetic material, for example, justify public access to the genetic information? It is easy to see why the answer to this question has to be "no." Consider a hypothetical case.

Emma loves Carl and he reciprocates her love. He begs her to give him a lock of her lovely hair and she does so. Unfortunately, a short while later they fall out. Emma falls in love with Fred; Carl is wild with jealousy and decides to take revenge. He remembers the lock of hair and in his anger sends it off to a laboratory for genetic analysis. Once there, the DNA is isolated from Emma's hair and analyzed. It turns out that Emma is prone to a serious disease; Carl insidiously informs Emma's health insurance company, which in turn cancels her policy.

This story definitely places Carl's character in a bad light. Did he also *act* in a way he should not have? Our intuition tells us that Carl's deed was immoral – and also unlawful. But why? We are certainly unable to justify our intuitively negative verdict of Carl's deed in terms of "body" and "ownership." Carl violated neither Emma's right to freedom from bodily injury (she cut off the lock of hair herself), nor her right of ownership (Emma gave Carl the lock of hair). One could point out that, although Emma gave away her lock of hair and, with it, the genetic matter it held, it was not her intention to give Carl access to genetic information about her person. This is applicable insofar as Emma's intentions are concerned, but it does not include a conclusive reason why Carl was not justified in accessing the information contained within his present. If Emma had given Carl a record, she would not have been able to maintain after their separation that she had only given him a sound-carrier to look at, and that Carl had no right to listen to the record. To put it succinctly: there is no plausible reason, in terms of ownership alone, why there should be a difference between a record and a lock of hair. Why should Carl have the right to access the acoustic information on the record (if it belongs to him), yet not have the right to access the genetic information in the lock of hair (if it belongs to him)? Or to put it another way: owning genetic material is not a justification for accessing and using the genetic information it holds. This becomes especially clear if we take a look at a second variation of our example.

This time Carl and Emma remain in love – for the rest of their lives. Yet Carl has never liked his father-in-law, and his hatred grows by the day. Once again, he sends off Emma's lock of hair to a gene technological laboratory. There it is established that Emma's father has a predisposition for a grave disease (e.g., Chorea Huntington). Once again, Carl sends this information to those whom it could interest, this time in order to inflict harm upon his hated father-in-law.

This second case makes the problems raised by the attempt to explain the normative status of the human genome in terms of "freedom from bodily injury" and "ownership" very clear. In diagnoses and analyses, it is the genetic information and not the genetic matter which is crucial, and information is by its very nature "fleeting." Genes also extend beyond the individual body and may thus be seen in terms of "shared ownership." This idea has been considered by a working group in Great Britain:

Because of the nature of genes, it may be argued that genetic information about any individual should not be regarded as personal to that individual, but as the common property of other people who may share those genes, and who need the information in order to find out their own genetic constitution. If so, an individual's prima facie right to confidentiality and privacy might be regarded as overridden by the rights of others to have access to information about themselves, rather as rights to privacy in the home presumably do not extend to denying access to other people whose property is being held on the premises (Working Group of the Royal College of Physicians, 1991, p. 410).

The idea of "shared ownership" has not yet found a place in other European countries, although it seems to be a corollary of the superindividual nature of the genome, and although it offers an argument for banning interventions in the germ-line, human-animal hybrids, and the cloning of human beings. If the genome is collectively owned, then manipulations of this genome can be legitimate only if all the co-owners have given their consent; but this is obviously impossible, due to the many future persons who are (or may be) amongst these co-owners.

To explain this, we have to turn to the second reason why the right to freedom from bodily injury and the right to ownership play only a minor role in the European debates on the normative status of the human genome. There is a deep and fundamental reluctance to accept the human being as the "owner" of his or her body, or to approve free disposition of this body – even by this "owner." In some European countries, like France and Great Britain, the individual is not legally considered to be the owner of his/her body; in other countries, like Germany and the Netherlands he/she is limited in his/her rights of disposition. This is also true of parts of the human body. The limits to these rights of disposition may be seen, for example, in wide-ranging restrictions upon all forms of commercialization of human organs and parts of the body, including blood. In parts of Europe, all forms of dealing with human organs are banned. In Germany, this is not the case, although the selling of organs is deemed to be "sittenwidrig" (immoral), meaning that contracts which have been drawn up to this effect will not hold up in Court. It was therefore easy for the countries involved in the above mentioned Bioethics Convention to reach a consensus that the human body, including its individual parts, may not be a source of profit. Article 11 of the Convention declares, "The human body and its parts shall not, as such, give rise to financial gain." The Draft Explanatory Report, added to the text of the Convention, justifies this prohibition with "the principle of human dignity, set forth in the preamble and in Article 1" of the Convention. Legislation against any kind of commercialization of human organs and body parts is currently being prepared on a European level.

What seems to be important here is that the normative status attributed to the human body and its parts is not only there in order to prevent third-party interventions, it also imposes certain obligations on the individual in question. This is obviously based on views about the morally correct and incorrect ways of treating the human body, which is not to be disposed of with absolute freedom by *any* of those involved. The human body is not seen as an object at anyone's free disposal; this is where the reluctance to discuss the moral or legal legitimacy of biotechnological options in term of ownership stems from. To put it briefly, the normative status of the human body does not only ensure the self-determination of the individual over his/her body, but at the same time inflicts restrictions upon that individual.

On the one hand, the right to freedom from bodily injury and the right to ownership guarantee the human genome a normative status; but they do this in an incomplete and unsatisfactory way. Both of these rights or principles are simultaneously too narrow and too wide. They are too narrow because they only protect genetic material and only cover what is of main importance: the genetic information, albeit indirectly, to the extent with which it is embodied within the genetic material. They are too wide because neither of these rights is restrictive enough to match the widespread consensus about the correct and incorrect ways of treating the human body. Discussing the normative status of the human genome in terms of ownership would imply that the "owners" have a right to unrestricted disposition of their "genetic property," which – according to this widespread consensus – they must not have.

III. PERSON(ALITY) AND HUMAN DIGNITY

The normative status of the human genome is usually justified, not only with regard to interventions, but also with regard to diagnostic insights into it, by reference to the terms "person(ality)" and "human dignity." Both concepts have an important function within the debate in Germany, bridging law and morality. They are closely connected, even if this relationship is somewhat unclear. The German *Basic Law* states "[e]veryone has the right to unfold his personality" (art. 2.1) and the German Federal Constitutional Court has repeatedly stressed that the Constitution, by placing the freedom of human personality on the highest rung of the ladder of values, recognizes its unique value, its unique status. And article 3 of the German *Basic Law* states that the rights attached to personal status are independent of sex, racial origin, nationality, religious and political beliefs, i.e., they apply to all human beings. This also includes human beings who are not in a position, whether temporarily or permanently, to exhibit specifically human capabilities (such as rational thinking), including embryos. There is little doubt that this right freely to unfold one's personality includes the right of each individual to determine over the acquisition and use of his/her personal data. An important decision of the German Federal Constitutional Court from 1983 confirmed this as a basic right to informational self-determination. Although this decision referred to "conventional" data, there can be no reasonable doubt that genetic information – as a particularly sensitive type of personal data – is party to this protection. The normative status of genetic information thus appears as a necessary corollary from the right freely to unfold one's personality, and the right to informational self-determination, which may be deduced from it. The normative status of the human genome thus receives added support from the Constitutional right to unfold one's personality freely, not only third-party interventions, but also insights into the genome may be prevented by appealing to this right. It is important that this principle covers not only genetic material but, more specifically, genetic information, the protection of which – as we have already seen – is at stake.

Yet the question remains of whether one may deduce from this right an unlimited disposition of the genome on the part of the individual in question. Does the right to unfold one's personality freely automatically lead to such unrestrained disposition? Under German Law, this is not the case. Like the right to ownership, the right to unfold one's personality freely is also subject to certain restrictions. Besides the obvious restriction that, in freely unfolding one's personality, one has to respect the rights of others, there are further restrictions resulting from what, in German legal-speak, is termed "Sittengesetz" (moral law). According to a definition by the German Supreme Court, these are "fundamental notions of the moral behavior which is demanded of the individual by reigning legal and cultural views within the social community" (Entscheidungen des Bundesgerichtshofs, p. 50).

Without wanting to examine the background and implications of this legal concept any further here, it should be noted that it infers certain restrictions concerning the treatment of one's own body. According to §226a of the German Criminal Code, bodily injury is punishable even if the receiver gives his/her consent, insofar as this consent conflicts with the moral law. An individual who has his/her leg amputated only in order to unfold his/her personality may not be punishable, but a physician carrying out such an operation without evidence of any medical indication certainly is. For such an amputation to be permissible, *medical reasons* are required. Applied to the (individual) human genome, this means that interventions are not subject to restriction as long as there is a medical reason for them. This is in complete accordance with articles 16 and 17 of the European *Bioethics Convention.*

> An intervention in the human genome may only be undertaken for preventive, therapeutic or diagnostic purposes and as long as the aim is not to interfere with the germ cell line . . . Tests which are predictive of genetic diseases or that may identify a genetic predisposition to a disease may only be performed for health purposes or for scientific research linked to health purposes.

The goals of these articles are to exclude any intervention aimed at modifying genetic characteristics not related to a disease, and to prohibit any intended interference with the germ cell line. Of prime interest here is the implication that the wishes of the individual in question, and his or her consent to the genetic intervention, are seen as a necessary but not sufficient condition for the intervention to be legitimate. Certain biotechnological options which will possibly become feasible in the future are therefore prohibited, even if they might be viewed by certain individuals as part of unfolding their personality freely. They include, for example, genetic interventions in order to increase one's performance (genetic doping) or to retain a continual level of endorphin production, to the extent that alcohol or drugs become superfluous. At the present time, the signs are that these kinds of self-realization will not be permitted in Europe. It remains to be seen, however, exactly where the line will be drawn between what counts as legitimate self-realization and what counts as abuse of gene technology.

Although the concept of person or personality plays an important role within the ethical and legal debate on the normative status of the human genome, it is the concept of human dignity which must be seen as the key category within this debate. In Germany, it has been included in the *Basic Law* more than anything as a reaction to the traumatic experience of Fascism. Article 2, §1 declares it to be a leading political and moral concept. "Human dignity is untouchable. It is the duty of the State to respect it and protect it." There is broad consensus that the concept of human dignity has a decisive function in the legal and moral evaluation of gene technology. Wolf-Michael Catenhusen, chairman of the German Parliamentary *Enquete-Kommission Gentechnologie* from 1985–1987, formulated it as follows.

> The concept of human dignity is especially suitable as a basis for deciding the matter of how far we are allowed to go in applying gene technology to human beings. Yet in Germany we are particularly sensitive with regard to the application of science and technology to human beings because of our experiences with the Third Reich. Human dignity leads to the rejection of positive eugenics, the attempt to improve the human being genetically, to breed the human being. Our experiences with the Third Reich prevent us from approaching this question as freely as in Great Britain or the U.S.A., for example. Germ-line therapy and prenatal diagnostics using gene technology also infringe human dignity (Catenhusen, 1988, p. 40).

The key position held by the concept of human dignity within the debate on applying gene technology to human beings is by no means restricted to Germany. Whilst it is seldom referred to in the Anglo-Saxon countries,[2] it is, for example, very important in Switzerland. It is a key notion in the Bioethics Convention, too, which declares as its purpose and goal in article 1:

> Parties to this Convention shall protect the dignity and identity of all human beings and guarantee everyone, without discrimination, respect for their integrity and other rights and fundamental freedoms with regard to the application of biology and medicine.

Even though there is broad consensus about the key role which the concept of human dignity plays, there is no consensus about what exactly is meant by "human dignity." This is the case not only for its content, but also for its implications with regard to the problems presented by gene technology. It is possible to distinguish two different directions of interpretation of "human dignity" within this context, one of which I have called elsewhere "substantialist," the other "subjectivist" (Bayertz, 1994).

The substantialist interpretation assumes more or less fixed ideas about how the human being is created and how it must behave in order to possess dignity. The German Federal Constitutional Court has often spoken of the "Menschenbild des Grundgesetzes" (constitutional image of man) in this context, and has incorporated it into the constitutional law. This interpretation is especially attractive for those who see gene technology as a threat to the human being and human dignity; it increases the chances of limiting human self-determination and self-alteration and may be taken as the basis for a categorical prohibition of all interventions in the human genome. But what are the elements of such a "substantial" view of the human being? Two criteria may be mentioned here.

(a) Human dignity is connected more or less closely with the "naturalness" and spontaneity of human procreation. In the report of the *Enquete Kommission "Gentechnologie"* of the German Bundestag we find the following statement: "[a]ny evaluation must follow from the assumption that the core of humanity is based on natural becoming, not on technical manufacture, and not on a social act of recognition. Human dignity has its essential foundations in birth and the naturalness of human origins, shared by all human beings. The human being enters the world as a 'begotten' and new-born-member of society. It is only 'made' in a very superficial sense. The parents may not, however, control how their genes are combined in their child's genome. One human being's physical and mental make-up is not due to the planning or arbitrariness of other human beings. The human being is a 'product of chance' " (Bericht der Enquete-Kommission, 1987, p. 187). It should be emphasized, however, that this passage does not reflect the opinion of the majority of the committee. Very similarly, the European Parliament declared back in 1982 that the rights to life and human dignity protected in articles 2 and 3 of the European Convention for Human Rights include the right to a genome which has not been artificially manipulated (Europäisches Parlament, p. 12).

(b) Gene technology is seen by many as a threat to the openness of individual human life and the resulting necessity of self-discovery. In a much discussed speech, Ernst Benda, former German Minister for Home Affairs, and President of the German Federal Constitutional Court, declared the "incompleteness" of the human being to be the core of its being and its dignity.

To put it pointedly, it is a question of whether it is really the human spirit, dividing it from the rest of Nature, really the human ability to make moral, responsible decisions which make up the core of its being, and not its incompleteness and inadequacy. This inadequacy is the result of a comparison with the perfection to be found in animals, ideally adapted for their – albeit limited – purpose within Nature ... If, with gene surgical interventions, scientific research succeeded in making the process of striving towards somewhere near the ideal, however good this intention may be, superfluous, a process which is always accompanied by the risk of failure, then we would have a new being which would only be physically similar to the human being with which we are familiar. This would no longer be the human being referred to in the Constitution because, in losing the necessity of taking sole responsibility for shaping its own life, the human being would also lose the ability to do so ... Science will never cease to examine the question of human nature. Yet today it is necessary to redefine the limits beyond which interventions are inadmissible for changing the essence of the human being. The legal system to be set up by the State is thus faced with a decision of enormous proportion. The State will not be able to escape its responsibility to maintain human dignity (Kaufmann, 1991).

Similarly, the commercialization of the human body and its parts is rejected by the European Bioethics Convention with reference to the principle of human dignity. Reference to human dignity serves here to restrict, at least under certain conditions, the freedom of the individual to decide and act as he or she pleases. These attempts to interpret the concept of human dignity from a substantialist point of view obviously give rise to a whole series of questions, as well as provoking a number of objections. In conclusion, I would like to point out that the extensive use of the term "human dignity" in connection with the problems involved in gene and reproduction technology has not escaped criticism within philosophical (Braun *et al.*, 1987; Catenhusen, 1988) and legal (Graf, 1987; Neumann, 1988) literature. The substantialist interpretation of the concept of human dignity amounts to making a particular view of humanity legally binding, and is thus hardly compatible with the pluralism of world views and life forms existing within modern societies, as with the philosophical and ideological neutrality of the Law.

The "subjectivist" interpretation of human dignity avoids any essentialist view of humanity and human nature. It assumes that self-determination is the essence of the human being. The historical locus classicus of this implication from a philosophical point of view is Kant's second formulation of his categorical imperative: "[a]ct in such a way that you always treat humanity, whether in the form of your own person or in that of another, as an end, and never just as the means to an end" (1968, p. 429). With regard to the debate on the normative status of the human genome, this interpretation is obviously not an obstacle, at least not a principal one, to intervening in the human genome – as long as these interventions find the agreement of all involved individuals. It is even possible to regard far-reaching projects of biotechnological human self-alteration as a new stage in human independence from evolutionary chance, as a realization of a higher level of self-determination, and therefore as an expression of human dignity. Whereas the substantialist interpretation tends to render human dignity independent of its carriers as a separable, objective value, and, as a consequence, not only tries to protect it from third-party harm, but also from self-inflicted harm, the subjectivist interpretation dissolves human dignity in individual autonomy. Any restriction to this autonomy appears not as a protection of human dignity, but as its violation.

In the moral and legal debates taking place, the substantialist and subjectivist interpretations are not strictly separated. Mixed versions with a tendency to one side seem to be preferred. The "substantialists" are generous when it comes to individual self-determination; the subjectivists concede that certain conceivable biotechnological projects may violate human dignity. (One example would be the breeding of an inferior race to carry out mundane, menial tasks.) The concept of human dignity thus, on the one hand, represents a medium within which concrete, argumentative bargaining processes concerning the limits of the normatively permissible may occur. On the other hand, using it in order to put an end to this process of argumentative bargaining is demanding too much. "But human dignity is only a formula used to put a name to an existing consensus. It is not an argument which may be used to create a non-existent consensus" (Kaufmann, 1991, p. 21).

IV. SUMMARY AND CONCLUSION

First, all of the issues dealt with in this paper are highly controversial elements of the philosophical and public debate in Europe. Nevertheless, some converging tendencies are shining through on a politico-legal level. I have referred to these converging tendencies as a specifically European "moral culture." Characteristic of this moral culture is the attempt to conserve a certain moral content, refuting all the erosions of modernity. Obviously enough, this moral content has its roots in basic convictions of the Christian tradition.

Second, part of this moral culture is a certain normative image of the human being and its integrity. This includes the view that the human body is more than a "piece of nature," is more than a mere object at its "owner's" disposal. The normative status of the human body imposes certain obligations upon everybody, including the individual in question. Although there is a wide range of legitimate "uses" of the human body, a broad consensus exists that at least some kinds of "uses" and treatments are morally and legally wrong. This holds for the human genome as well.

Third, the right to unfold one's personality and the informed consent given by an individual to any kind of technological intervention in the human genome are, therefore, seen as a necessary but not sufficient condition for this intervention to be legitimate. There have to be good reasons for such interventions, i.e., reasons which are plausible to others and which are compatible with the normative image of the human being. Medical reasons are always accepted as such good reasons – at least when medicine is aimed at restoring this image of the human being. This is why, to mention another example, interventions in human reproduction like AID or IVF are usually justified medically: as therapy for unwanted childlessness, and not – as in the United States – as a means of free choice of reproductive options.

Department of Philosophy
University of Münster
Münster, Germany

NOTES

[1] Translated into English by Sarah L. Kirkby.

[2] One example is W. French Anderson: "[u]nder what circumstances would human genetic engineering not be a moral good? In the broadest sense, when it detracts from, rather than contributes to, the dignity of man. Whether viewed from a theological perspective or a secular humanist one, the justification for drawing a line is founded on the argument that, beyond that line, human values that our society considers important for the dignity of man would be significantly threatened" (Anderson, 1989, p. 685f).

BIBLIOGRAPHY

Anderson W.F.: 1989, 'Human gene therapy: Why draw a line?', *The Journal of Medicine and Philosophy* **14**, 685.

Bayertz K.: 1994, *GenEthics; Technological Intervention in Human Reproduction as a Philosophical Problem*, Cambridge University Press, Cambridge.

Benda E.: 1985, 'Erprobung der Menschenwürde am Beispiel der Humangenetik', *Aus Politik und Zeitgeschichte, Beilage zur Wochenzeitung "Das Parlament"*, B. 3/5.

Bericht der Enquete-Kommission: 1987, 'Chancen und Risiken der Gentechnologie', *Bundestagsdrucksache* 19/6775, Bonn. January 6.

Birnbacher D.: 1987, 'Gefährdet die moderne Reproduktionsmedizin die menschliche Würde?', in V. Braun *et al.* (eds.), *Ethische und rechtliche Fragen der Gentechnologie und der Reproduktionsmedizin*, München, pp. 77–88.

Catenhusen W.M.: 1988, 'Kodifizierung der Ethik am Beispiel der Gentechnologie', in Gesellschaft Gesundheit und Forschung (eds.), *Ethik und Gentechnologie*, Frankfurt/Main, p. 40.

Council of Europe, Steering Committee on Bioethics: June 30, 1994, *Preliminary Draft Convention for the Protection of Human Rights and Dignity of the Human Being with regard to the Application of Biology and Medicine: Bioethics Convention*, Strasbourg.

Entscheidungen des Bundesgerichtshofs 67, p. 50.

Europäisches Parlament: 1982, Empfehlung 934 Betr. die Gen-Manipulation, Bundestagsdrucksache 9/1373, S. pp. 11–13.

Graf Vitzthum W.: 1987, 'Gentechnologie und Menschenwürdeargument', *Zeitschrift für Rechtspolitik* **20**, 33–37.

Kant I.: 1968, *Grundlegung zur Metaphysik der Sitten*, Akademie-Textausgabe, Bd. IV, de Gruyter, Berlin.

Kaufmann A.: 1991, 'Rechtsphilosophische Reflexionen über Biotechnologie und Bioethik an der Schwelle zum dritten Jahrtausend', *Archiv für Rechts- und Sozialphilosophie*, Beiheft.

Mauron A. and Thévoz J.-M.: 1991, 'Germ-line engineering: A few European voices', *The Journal of Medicine and Philosophy* **16**, 649–666.

Neumann U.: 1988, 'Die "Würde des Menschen" in der Diskussion um Gentechnologie und Befruchtungstechnologien', *Archiv für Rechts- und Sozialphilosophie*, Beiheft **33**, 139–152.

Schweitzer, J.: 1985, *In-vitro-Fertilisation, Genomanalyse und Gentherapie. Bericht der gemeinsamen Arbeitsgruppe des Bundesministers für Forschung und Technologie und des Bundesministers der Justiz*, Verlag, München, 1985.

Schöne-Seifert B.: 1990, 'Philosophische Überlegungen zu "Menschenwürde" und Fortpflanzungs-Medizin', *Zeitschrift für philosophische Forschung* **44**, 442–473.

Working Group of the Royal College of Physicians Committees on Ethical Issues in Medicine and Clinical Genetics: 1991, *Ethical Issues in Clinical Genetics*, RCP, London.

H. TRISTRAM ENGELHARDT, JR

MORAL PUZZLES CONCERNING THE HUMAN GENOME: WESTERN TABOOS, INTUITIONS, AND BELIEFS AT THE END OF THE CHRISTIAN ERA

I. INTRODUCTION

In his excellent study of genetic engineering and interventions into human reproduction, Kurt Bayertz explores the European attempt to maintain a "moral culture" in the area of genetic engineering and artificial reproduction. As Bayertz indicates, this concern is particularly salient in Germany. No doubt this circumstance is significantly determined by recent German history. But, as he notes, such appeals exist in Europe outside of Germany. In providing his analysis of this very particular cultural and moral phenomenon, Bayertz indirectly offers an account of the stress lines engendered by a profound change in European culture: its significant deChristianization.

This change has been underway in the West since at least the 16th century and has been expressed in the project of securing the moral commitments of Western Christianity by an appeal to reason rather than by a confession of belief. There has been a commitment to secure the contents of Western Christianity outside of a Christian community and without acknowledgement of the Christian God. One might think, for example, of how Immanuel Kant appeals to reason itself to secure a morality for all and then finds much of the particular content of a Christian morality in reason. If one names this attempt the project of modernity or the Enlightenment, and then identifies post-modernity as the recognition of its failure, that is, the failure to secure a universal moral narrative through appeal to reason itself or to a secularly disclosable canonical view of human nature, one can then better understand the European project of securing a moral culture. It is an attempt to sustain the particular moral commitments of West Europeans without either

Kazumasa Hoshino (ed.), Japanese and Western Bioethics, 181–186.

recognizing their particularity or re-embedding them in the religious faiths from which most of them sprang.

As I have argued elsewhere, this attempt to secure the content of European morality by an appeal to general secular reason cannot succeed (Engelhardt, 1991; 1996). Here I will content myself with (1) taking stock of the appeal to a moral culture as a means of evaluating the morality of enhancing human abilities through genetic engineering, (2) connecting Bayertz' concerns with a "moral culture" to Alasdair MacIntyre's account of taboos in Hawaii and Europe, and (3) then advancing some reflections regarding the contemporary character of Western bioethics. A point of warning in all of this is in order. Though I speak of the West entering a post-Christian era, I advance this as a cultural assessment, not a normative claim. I am, after all, a Christian.

II. MORAL CULTURE AND THE OPPOSITION TO ENHANCING HUMAN CAPACITIES THROUGH GENETIC ENGINEERING

The set of moral values that constitutes the moral culture to which Bayertz turns his attention frames a content-full understanding of how humans should protect both embryos and the contemporary character of the human genome. Importantly, the moral culture that Europeans in general and Germans in particular wish to sustain is opposed to the use of genetic engineering when it steps beyond curing diseases and is used to enhance human abilities.

There is no question that serious concerns regarding the safety of germ-line genetic engineering are in order. However, the concerns that can be articulated in general secular terms, open to all and without special appeal to particular religious and cultural understandings, are primarily prudential or focus on protecting actual persons from being used without their actual consent. Before one intervenes to change the human genome in order not simply to cure diseases, but to enhance human abilities, it is prudent to be fairly sure that success is likely without unforeseen risks to the persons treated and to future offspring. Such prudential concerns are substantial, significant, and understandable. But beyond these and the canons of competent consent, little else seems securable without an appeal to a very particular moral vision. Hence, one finds the invocation of a "moral culture."

It is impossible to justify an objection in principle against the use of germ-line engineering to cure genetic diseases in a definitive fashion outside of a special appeal to a very particular moral understanding. If, for instance, one were able not only to cure retinoblastoma, but to remove the gene and replace it with a normal one so as to protect not just the person treated but all future offspring, how could that be anything but a preferable outcome? Again, prudential issues will surely be significant. Concerns not to do more harm than benefit are always in order. But if these concerns can in principle be met, then such interventions should be allowed.

Once one has gone this far in accepting germline genetic engineering, the difficulty is then in establishing why one may not also cautiously move to enhance human functions and abilities, not simply cure diseases. What if one could engineer away some of the risks of coronary artery disease to which humans are generally exposed? What if one could through genetic engineering lower the risk of prostatic carcinoma, as well as osteoporosis associated with menopause? What would in principle be wrong with cautiously taking steps towards enhancing human physical and psychological abilities? Again, prudential concerns would surely be significant, as well as the need to have the consent of the actual persons directly involved. But beyond such concerns, what could be said in general secular terms to show the moral impermissibility of such interventions without an appeal to a particular set of religious or cultural commitments? To have objections in principle, one must possess a normative vision of the current human condition. But how is one to secure this vision with which to prohibit in general secular terms the use of genetic germline genetic engineering to enhance human abilities?

III. EUROPEANS AND THE POST-CHRISTIAN AGE

The struggle in Europe, and in Germany in particular, to maintain a moral culture can best be understood as a secular attempt to maintain elements of the Christian culture that fashioned Western Europe. The idea that one might be able to maintain elements of that culture through quasi-philosophical appeals may appear especially plausible in the context of European culture. Western Christian culture was in great measure committed not simply to accepting God's revelation, but to holding that much of Christian morality could in fact be understood and disclosed by right reason. The correct way of acting, the *recta ratio agibilis*, depended on the disclosure of a *recta ratio*, a right reason. The difficulty is that the hope to secure a contextless, general reason that can disclose the content of a particular religion and culture is vain. The failure in Europe to articulate general secular rational grounds for its moral culture is a function of this inability. Yet, there are cultural reasons for not being able easily to accept this inability. Its acceptance would mean acknowledging the extent to which Europe has entered a post-Christian era. This would also mean acknowledging that the Christianity of the West had committed Europe to a view of reason's ability that could not be sustained by secular reason. This view of Christianity, it might be observed, has led to a crisis not only for European secular culture, but for Western Christianity itself (Buckley, 1987).

In his *Three Rival Versions of Moral Enquiry*, MacIntyre observes how in the 19th century many of the moral intuitions and commitments of Hawaiians appeared both to Hawaiians and to Europeans as mere taboos, as sentiments of moral disapproval disarticulated from intact supporting moral traditions and social practices (MacIntyre, 1990). Both Hawaiians and Europeans could recognize moral rules and intuitions that had grown out of the traditional Hawaiian culture, as having an arbitrary character. Both Europeans and Hawaiians could recognize the extent to which Hawaiians had stepped away from their traditional culture, which had framed and sustained those moral rules and intuitions. Isolated from those moral traditions and the moral practices they carried with them, these rules and intuitions were at best disembodied insights that could guide human behavior only as long as their general rationality was not challenged. MacIntyre then observes that what was occurring with regard to Hawaiians was also taking place with the Europeans.

Western Europeans have been left with shards and pieces of what once was a coherent and encompassing morality. There is often the sense that particular undertakings are wrong, but general secular arguments cannot establish their wrongness. Nor is their place in a larger framework of commitments any longer obvious. There is a disconnection from the religious context within which such wrongness was once appreciated. This circumstance frames the difficulty in giving a secular philosophical defense of the particular content-full intuitions and rules of Western European morality apart from the traditions that fashioned them. The impossibility of that task with respect to Hawaiian culture led to the collapse of traditional Hawaiian taboos. In *Three Rival Versions of*

Moral Enquiry, as well as elsewhere, Alasdair MacIntyre suggests the same is occurring in Europe (MacIntyre, 1981; 1988; 1990).

IV. WESTERN AND JAPANESE BIOETHICS: MORAL CULTURES IN THE PLURAL

Bayertz contributes significantly to the comparison of Western and Japanese bioethics. He does this by raising the question of the origin and justification of the moral content that particular bioethics endorse. Because of its combination of rationalist and universalist aspirations, the West committed itself to securing the particularity of its claims by reason alone. It did not look to the particularity of faith, culture, or history. Indeed, it has in great measure been blind to the particularity of its content-full morality. Nor has it recognized how that morality has shattered into numerous competing moralities, thus not appreciating the plural character of bioethics even in a secular context. In addition, by attempting to appeal to general rational argument rather than to established traditions and practices, Western societies have disconnected their traditional moralities from their supporting structures. This has rendered much of their substance into unsecured moral intuitions, which can then appear at best as mere taboos.

All of this has occurred without European bioethics frankly confronting the particularity of its commitments or the precariousness of its content. European societies still yearn to maintain much of the content of their traditional moralities, the substance of which, at least in many areas, can appear only like the taboos of the Hawaiians, once it is separated from its framing traditions and metaphysical supports. Bayertz's account of the European, and in particular German, attempt to maintain a moral culture is both an example of one of the products of that yearning as well as an illustration of how cultural content shapes a particular bioethics. Europe's moral culture is of course only one morality over against other alternative moralities, and not the one canonical rational morality. That is, it has content that cannot be secured through sound rational argument. The same can be said regarding Japanese bioethics, though Japanese bioethics is not set within a background culture with such strong expectations from reason nor claims to generality of governance. By illustrating the particularity of European bioethics, Bayertz contributes importantly to the recognition of the plural character of bioethics and to the appreciation of the different dynamics that shape Western and Japanese bioethics.

Center for Medical Ethics and Health Care Policy
Baylor College of Medicine/Rice University
Houston, Texas, USA

BIBLIOGRAPHY

Buckley, M.J.: 1987, *At the Origins of Modern Atheism*, Yale University Press, New Haven, Conn.

Engelhardt, H.T., Jr.: 1996, *The Foundations of Bioethics*, 2nd edition, Oxford University Press, New York.

Engelhardt, H.T., Jr.: 1991, *Bioethics and Secular Humanism*, Trinity Press International, Philadelphia.

MacIntyre, A.: 1981, *After Virtue*, University of Notre Dame Press, Notre Dame, Ind.

MacIntyre, A.: 1988, *Whose Justice? Which Rationality?* University of Notre Dame Press, Notre Dame, Ind.

MacIntyre, A.: 1990, *Three Rival Versions of Moral Enquiry*, University of Notre Dame Press, Notre Dame, Ind.

PART FIVE

CROSS CULTURAL DIVERSITY AND POST-TRADITIONAL MORALITY

RUIPING FAN

THREE LEVELS OF PROBLEMS IN CROSS-CULTURAL EXPLORATIONS OF BIOETHICS: A METHODOLOGICAL APPROACH

I. INTRODUCTION

It is believed that cross-cultural explorations of ethics in general and bioethics in particular are important in disclosing apparently different moral standards and moral practices among distinct cultures. On the other hand, cross-cultural comparisons of morality are often made by ethicists in order to demonstrate the universal validity of certain basic (or core) moral standards in every culture, no matter how divergent cultures appear to be. In this regard, cross-cultural examinations of bioethics are an extension of standard Western bioethics, "the family of secular approaches rooted in the theories and principles of analytical moral philosophy that are dominant in the English-speaking world" (McKenny, 1996, p. 74). One representative and very influential standard Western bioethical theory claims itself to be a common-morality theory which "takes its basic premises directly from the morality shared in common by the members of a society – that is, unphilosophical common sense and tradition" (Beauchamp and Childress, 1994, p. 100). Accordingly, many believe that most fundamental secular moral premises and principles thus obtained from a common sense of Western morality apply universally in the West. They are supposed to be basic secular moral beliefs shared by everyone in spite of diverse religious, ideological, or other commitments. In this regard, standard Western bioethics is considered to have successfully transcended the diversity of moral communities which marks pluralist Western society.

When standard Western bioethicists encounter a non-Western culture, they are tempted to extend the supposed validity of their general secular moral principles to the alien culture. Presumably, they can offer an argument as follows. Since standard Western bioethics is derived

Kazumasa Hoshino (ed.), Japanese and Western Bioethics, 189–199.

from "the morality shared in common by the members of [Western] society" and is held to transcend the divergent religious and metaphysical commitments of multiple moral communities of pluralist Western society, standard Western bioethics should similarly apply to a non-Western society in which, as in Western societies, distinct particular moral communities exist.

As to whether there is such a thing as a "morality shared in common by the members of [Western] society" and captured universally by the basic principles of standard Western bioethics is a question beyond the scope of my concern in this essay. Here, I will focus my attention only on the following methodological issues: first, even if it is the case that there is a "morality shared in common by the members of [Western] society" and the basic premises and the fundamental principles of standard Western bioethics are directly taken from it, we are still yet to establish whether there is a morality shared in common by the members of a non-Western society; and, second, even if there is a morality shared in common by the members of a non-Western society, we still do not know whether it is the same morality as that shared by members of Western society.

As a preliminary point, I would like to indicate an inappropriate method which is often used in cross-cultural studies. In comparing morality between two cultures, Western bioethicists tend to frame their arguments by identifying certain cases that are found in a foreign culture and which appear to presuppose a set of moral principles to which these Western bioethicists have already committed themselves. In this way, they match their own moral principles and theories with certain cases and facts found in a foreign culture and thereby "successfully" confirm the cross-cultural objectivity of their own moral beliefs and the communality of moral understandings between the two cultures, without even referring to any particular moral premises and principles that the foreign people actually hold. Thus, they believe they have won a victory over a particular version of moral relativism which holds that moral understandings and principles are culturally relative.

However important it is to defeat moral relativism, such a facile victory is made at the price of endorsing a methodologically inappropriate approach. Regardless of the fact that the moral understandings and principles employed in a foreign culture differ from those employed in the West, such bioethicists affirm the privilege of applying standard Western bioethical understandings and principles to the foreign culture without providing a justification of their governance. Indeed, it is a difficult matter to show that seemingly similar moral concerns and practices are actually equivalent, especially when they are set within quite different cultural contexts. Consequently, such explorations may involuntarily lead to adopting a moral dogmatism.

Again, I do not know whether eventually standard Western bioethics can be demonstrated to hold in non-Western societies. However, if mere dogmatism is to be avoided, it is at least methodologically necessary to justify applying standard Western bioethical principles to a foreign culture, when these principles differ from or conflict with the moral principles actually employed in the foreign culture. Accordingly, I suggest that there is the need for a serious assessment of culturally-based moral differences among cultures. If we fail carefully to address these differences in the first place, we will not be in a position to supply the needed justifications in order to identify cross-cultural bioethical principles, should they exist. I take this to be an unavoidable methodological requirement.

My methodological suggestions for cross-cultural bioethics are, as a result, modest. First, one must supply a justification for applying one's own moral principles to a foreign culture. Second, following the first suggestion, one must take moral differences among cultures seriously. Finally, amplifying the second suggestion, one must recognize the following three levels of issues at stake in a comparative study of the morality between two cultures.

At the first level, there is the issue of whether there are differences in the crucial moral vocabulary and fundamental moral principles employed within the two cultures. If such differences exist, there is the further problem of whether one system of moral vocabulary and principles can be adequately reduced to the other.

At the second level, there is the issue of whether there are differences in the very meaning of a moral principle that appears to be accepted by both cultures. If there are differences in meaning, there will be an additional problem of determining their theoretical origins and the practical implications of the differences.

Finally, at the third level, there is the problem of whether there are differences in rankings of moral principles or goods even if the same principles or goods are endorsed in both cultures. If there are differences in ranking, there will be a further problem of determining the bases for these differences in each culture as well as the practical challenge of assessing the implications of these differences.

Needless to say, focusing on differences at these three levels does not preclude recognizing significant moral agreements between two cultures. There certainly exist a great deal of moral agreements among cultures. However, differences are more important than agreements because it is differences rather than agreements that challenge belief in the universal validity of standard Western bioethics. By pursuing this methodological approach with careful attention to the issues at each of the three levels, we may eventually determine the extent to which standard Western bioethics admits of universal application.

In the following three sections, I will briefly address certain differences at these three levels between current Western culture on the one hand, and contemporary Japanese and Chinese cultures on the other. In doing so (1) I will not be able to offer an entire description and evaluation of all differences between these cultures; (2) although I will contrast Japanese and Chinese cultures with Western culture, I do not imply that Japanese and Chinese cultures are morally equivalent; (3) neither is it my intention to provide a comprehensive outline of morality for any of these three cultures. Rather, I will identify only several examples which I will argue are sufficient to establish the significance of this methodological approach to examining problems in cross-cultural explorations of bioethics.

II. DIFFERENCES IN MORAL VOCABULARY AND FUNDAMENTAL MORAL PRINCIPLES

Individual rights, liberty, autonomy, self-determination, equality, justice, fairness, etc. are integral to the moral vocabulary of modern Western people. These terms are at home in current Western society. However, though more and more Chinese and Japanese have come to understand this Western vocabulary, these are not the concepts generally used in the ordinary life of Japanese or Chinese people. In contrast, for the Japanese people, the key moral concepts include "Wa" (harmony) (Hoshino, 1995), "Amae" (dependence) (Doi, 1973), and "Taijisokuin" (great mercy) (Kimura, 1986). For the Chinese people, the crucial terms are "Jen" (humaneness), "Yi" (righteousness), "Li" (propriety), "Xiao" (filial piety), "Zhong" (faithfulness), "Ci" (benevolence), "Xin" (fidelity), "Xu" (reciprocity), "De" (virtue), and so on.

Similarly, while Western people employ such principles as respect for autonomy, beneficence, non-maleficence, and justice, the Chinese and Japanese people engage quite different ones. For instance, the five Confucian principles remain the cardinal moral principles that regulate the contemporary life of most Chinese. They are the principles of affection ("Qin") between parents and children, righteousness ("Yi") between monarch and subjects, function ("Bie") between men and women, order ("Xu") between the older and the younger, and fidelity ("Xin") between friend and friend (Mencius, 1970, p. 252).[1] These principles govern almost every aspect of Chinese moral life, the field of health care certainly included. If there is a common morality in Chinese society, these principles cannot be excluded.

Japanese and Chinese moral terms and principles are placed within theoretical moral systems which stand in contrast with that of standard Western bioethics. What is at stake is not whether the vocabulary or principles of one system cannot be translated into the language of the other system. The emphasis on an untranslatability between these different moral systems can be misleading. People can always devise a way to interpret alien concepts and principles in their own language. Neither is it the case that what is right according to one system is always wrong according to the other. There may be a great number of points of agreement between two moral systems with regard to judging particular cases. What is crucial, rather, is that *the standards by which one system determines how morally right-wrong distinctions are to be drawn are not the same across systems*. This difference between two moral systems often entails distinct ways of moral life and different moral judgments concerning certain particular cases. Accordingly, two such moral systems can be described most accurately as morally incommensurable, to recast the notion of "incommensurable" from the philosophy of science. Specifically, the system of Confucian ethics and the system of current Western ethics are incommensurable in the following three senses: (1) the moral standards employed are different between the two systems, (2) divergent moral practices are established under the two systems, and (3) disagreements between the two systems regarding particular cases cannot be resolved by appealing to a neutral standard, because no such neutral standard is available. In short, the incommensurability of the two systems leads to the predicament in which one system cannot be replaced by the other without a loss of moral content.

When standard Western bioethicists employ their own principles in Chinese or Japanese society, they substitute their own moral system for the Chinese or Japanese one. Is such a substitution proper? The answer depends upon whether it can be shown that the Chinese or Japanese bioethical system is less appropriate, less rational, or less true than the standard Western bioethical system. Alasdair MacIntyre has argued for a general way through which such matters can be addressed. He contends that one moral tradition can be recognized as inferior to a rival tradition if the following two conditions are satisfied (MacIntyre, 1988, pp. 361–365): first, the former tradition falls into a stage of "epistemological crisis" in which conflicts over rival answers to its key questions can no longer be settled rationally and by its own standards it ceases to make progress; secondly, at this tradition's stage of epistemological crisis, the alternative rival tradition is able to provide the resources to explain why it fails by its own standards and to provide the conceptual solutions to the crisis. In short, though MacIntyre does not think that there are neutral standards independent of the two rival traditions, he believes that by an appeal to the respective history of each tradition we may be able to judge the superiority of one over the other. The issue is that, even if MacIntyre is right, in order to make a judgment we must confront the challenge of providing a detailed and careful reconstruction of the histories of standard Western bioethics on the one hand, and Chinese or Japanese approaches to bioethics on the other.[2]

III. DIFFERENCES IN THE MEANING OF A MORAL PRINCIPLE

A number of moral principles can be used as examples of the challenges at this level of comparing moral principles. I will only take the principle of informed consent as illustrative in the following. As is well-known, in recent decades the principle of informed consent has become a vital element in Western medical practice. The principle requires physicians and researchers to disclose relevant information to patients or subjects so as to obtain their informed consent prior to any substantial intervention. Indeed, in standard Western bioethics, the basic sense of the principle is a requirement of first-person informed consent (Beauchamp, 1996).

As once in the West, so too in the traditional Chinese medical practice there was not an explicitly required principle of informed consent. The traditional principle of Jen (Humaneness, Love, or Virtue), however, does generally entail that physicians should explain sicknesses or diseases to the patient and obtain permission before performing any substantial treatment. However, informed consent as a specific moral principle did not appear in the Chinese medical literature until the 1980's. Even currently, the exercise of a principle of informed consent in Chinese medical practice expresses a feature of Chinese culture: familism. The Chinese assume that a family member is an integral element of the family and should be closely united with the rest of the family. This gives particular content to the principle of informed consent as shown in Chinese clinical practice. In contrast, while this principle in the West requires disclosing information to a patient himself or herself and obtaining consent directly, in China it requires disclosing information to a patient's family-representative and obtaining formal consent from the family-representative.

In fact, Chinese physicians usually confront the family as a whole rather than the patient in isolation, no matter whether treatment is performed in a patients' home as was the traditional practice, or in a modern hospital, as it occurs currently. The cultural assumption, which engages the active involvement of the whole family, is that a sick person should not sustain the burden of listening to a strange physician's explanations or be burdened with directly signing a consent form. A sick family-member is someone who should be taken care of by the rest of the family. Usually a representative of the family should arrange all matters with the physician.

This does not mean that a family-representative would make a decision without consulting the patient. To the contrary, insofar as is possible, the family-representative will consult not only the patient but also the other family members. The decision is made eventually by the family as a whole. Significantly, though the patient does not directly confront the physician and personally make an explicit contract for treatment, the decision cannot be made in conflict with the patient's wishes. Disagreements between the patient on the one hand and the rest of the family on the other, or among the rest of the family themselves, must be resolved before final permission is given to the physician. Chinese make a distinction between intra-familial and extra-familial authority.

As is shown by the other authors in this volume, an emphasis on self-determination in health care in the West has ancient religious roots as well as a modern Enlightenment background (Veatch, 1996; McKennny, 1996). Similarly, the feature of family-determination in medical care in China finds its theoretical and historical foundation in the Confucian understandings of the nature of family and of relations between individuals. Consequently, between the West and China there is a clear difference in the sense or meaming of the principle of informed consent.

There might be a sound philosophical argument that we have yet to put forward to show which sense of the principle of informed consent is correct, normal, or superior. But for the moment, at least one thing is clear: if there is such an argument, it cannot do its work without addressing background assumptions and theoretical understandings relating to each sense of the principle.

IV. DIFFERENCES IN THE RANKINGS OF MORAL GOODS AND PRINCIPLES

Not surprisingly, there are different rankings of moral goods and principles. Many individuals have experienced such ranking difference. Certainly, it is not the case that such difference occurs only cross-culturally. Even in the same culture, we encounter cases where there are no disagreements about which set of principles is relevant to the cases or about what meaning each principle carries, yet there are disagreements concerning what should be done in the cases (Brody, 1988, pp. 77–79). An explanation for such disagreements is that people have different orderings of the principles to which they appeal. Understandably, this kind of disagreement may occur more frequently between cultures.

As an ordinary example, we may notice an interesting difference in the ordering of certain goods that comparative psychology has repeatedly confirmed. Compared with their American counterparts, Chinese students tended to express a higher preference for Achievement, Deference, Order, Support, Nurturance, and Endurance, and a lower preference for Exhibition, Intraception, Dominance, Change, and Aggression (Yang, 1986, pp. 109–111).

In the field of medical care, both Chinese and Americans would accept the principle of respect for autonomy and the principle of beneficence as ethically relevant. They may all agree that a relevant sense of respecting patients' autonomy consists in physicians' "equipping them to overcome their sense of dependence and achieve as much control as possible" (Beauchamp and Childress, 1994, p. 127). They may as well agree that the principle of beneficence imposes an obligation on physicians to act for the benefit of their patients. Nevertheless, at least in many if not most cases, Americans would give more weight to this principle of respect for autonomy, whereas Chinese would give more weight to the principle of beneficence. Chinese would in addition likely acknowledge the importance of a principle of autonomy that expressed itself in affirming the role of families in determining treatment for their members.

It would be difficult to account for all the theoretical roots of such differences. But one element that authors often refer to may be worth mentioning here. While the Western theoretical tradition stresses the capacity that each individual person has to determine his or her own destiny, the Chinese Confucian tradition emphasizes the capacity that each individual person has to cooperate with other individual persons in determining his or her own destiny. Procrusteanly put, the Chinese culture is collective-oriented, whereas the Western culture is individual-oriented.

V. CONCLUSION

Based on the above discussions, I would like to draw two tentative conclusions. First, we are not in a position to answer the question as to whether standard Western bioethics will eventually be applicable to non-Western countries such as Japan or China. There is much yet to learn at this point. The crucial issue may not have much to do with the problem of moral relativism. Rather, the problem may concern how to deal with differences in moral focus and content in terms of (1) the reducibility or equivalence of moral vocabularies and principles, (2) the difference in the meaning of moral principles, and (3) the ordering of moral goods and principles.

Secondly, given that there are differences at these three levels, it is no surprise that moral controversies occur between Westerners and the Chinese or Japanese. Furthermore, their moral controversies cannot be resolved through sound rational argument or by an appeal to a jointly recognized world moral authority, because they do not share sufficient moral premises and there are no individuals or institutions that they mutually accept as being in authority to resolve moral controversies and to give content-full moral guidance. Accordingly, most Westerners and most Chinese or Japanese can be defined as moral strangers to each other.[3] Even when they appeal to the same set of moral principles, they may hold different rankings of the principles. When they seem to employ one moral principle, they in fact understand it in different senses.

More provocatively, they employ different and incommensurable moral vocabularies and assumptions. In this situation, the classical liberal assertion that John Stuart Mill termed "one very simple principle" is worth a re-emphasis: people should be left alone insofar as their actions do not harm others (Mill, 1986, p. 16). This formal principle should be instructive for us in resolving moral controversies at an international level. Everyone should have a right peaceably to pursue one's own view of the good life. And no one should be coerced to change one's peaceable moral vision. This ought to be the case not only intra-nationally but also internationally. Indeed, international moral controversies can best be settled through peaceable contracts, mutual agreements, and the mechanism of the free market.

Center for Medical Ethics and Health Policy
Baylor College of Medicine
Houston, Texas, U.S.A.

NOTES

I should like to thank H. Tristram Engelhardt, Jr., Gerald P. McKenny, and Mark J. Cherry for their critiques and suggestions concerning the first draft of this essay.

[1] For a collection of Confucian classics, see Confucius, 1971 and Mencius, 1970.

[2] In an essay concerning a comparative study of Confucian and Aristotelian virtues (MacIntyre, 1991), MacIntyre asserts that "Confucianism appears to face a recurrent type of dilemma: *either* it retains its highly specific and concrete character, thus tying itself to particular Chinese forms of social relationships of a traditional kind . . . *or* it makes itself relevant to types of social order in which these forms of social relationships do not or no longer exist . . . " (p. 120, original emphasis). However, I cannot see clearly where such a dilemma lies within the Confucian ethics. Since the issue is complicated, I will not address it here in detail. I will only point out the following issues. First, Confucianism does not lack a universal principle for human beings as such; the principle of Jen (humaneness, goodness, or benevolence) is just such a principle. Second, Confucianism does emphasize the particular five relationships (between parents and children, monarch and subjects, husband and wife, the older and the younger, and friend and friend). According to Confucianism, the respective natures of these relationships are such that the universal principle of Jen does require special obligations and virtues of people so as to engage successfully in these relationships. Finally, Confucians generally view these relationships as natural ones that are inevitably present in any appropriate society. They would disagree that there are moral types of social order "in which these forms of social relationships do not or no longer exist."

[3] For a systematic discussion of what it is to be moral strangers and moral friends, see Engelhardt, 1996, especially pp. 7–8.

BIBLIOGRAPHY

Beauchamp, T.L.: 1996, 'Comparative studies: Japan and America', in this volume, pp. 25–47.

Beauchamp, T.L. and Childress, J.F.: 1994, *Principles of Biomedical Ethics*, 4th edition, Oxford University Press, New York, Oxford.

Brody, B.: 1988, *Life and Death Decision Making*, Oxford University Press, New York, Oxford.

Confucius: 1971, *Confucian Analects, The Great Learning & The Doctrine of the Mean*, J. Leggs (translation with exegetical notes and dictionary of all characters), Dover Publications, Inc., New York.

Doi, T.: 1973, *The Anatomy of Dependence*, Kodansha International Ltd., Tokyo, New York & San Francisco.

Engelhardt, Jr., H.T.: 1996, *The Foundations of Bioethics*, second edition, Oxford University of Press, New York, Oxford.

Hoshino, K.: 1996, 'Bioethics in the light of Japanese sentiments', in this volume, pp. 13–23.

Kimura, R.: August 1986, 'In Japan, parents participate but doctors decide', *Hastings Center Report* **16**, 22–23.

MacIntyre, A.: 1988, *Whose Justice? Which Rationality*? University of Notre Dame Press, Notre Dame.

MacIntyre, A.: 1991, 'Incommensurability, truth, and the conversation between Confucians and Aristotelians about virtues', in E. Deutsch (ed.), *Culture and Modernity*, University of Hawaii Press, pp. 104–122.

McKenny, G.P.: 1996, 'Technology, authority and the loss of tradition: The roots of American bioethics in comparison with Japanese bioethics', in this volume, pp. 73–87.

Mencius: 1970, *The Works of Mencius*, J. Legge (translation with critical and exegetical notes, prolegomena, and copious indexes), Dover Publications, Inc., New York.

Mill, J.S.: 1986, *On Liberty*, Prometheus Books, Buffalo.

Veatch, R.M.: 1996, 'Authority and communitarianism: The ethics of terminal care in cross-cultural perspective', in this volume, pp. 119–130.

Yang, K: 1986, 'Chinese personality and its change', in M.H. Bond (ed.), *The Psychology of the Chinese People*, Oxford University Press, Hong Kong, Oxford, New York, pp. 106–170.

MARK J. CHERRY

MORAL STRANGERS: A HUMANITY THAT DOES NOT BIND

I. INTRODUCTION

Japanese and Western philosophers, physicians, and patients often meet as moral strangers. The essays in this volume underscore this circumstance. The essays explore particular bioethical issues such as human germ line therapy, autonomy, scarcity of medical resources, medical futility, and death and dying. Implicitly each underscores, intentionally or unintentionally, the deep sense in which we meet as moral strangers. We meet as individuals from diverse and often fragmented moral communities with diverse moral intuitions, premises, evaluations, and commitments, both inter- and intra-culturally (Engelhardt, 1996a). Medical clinical encounters, as well as attempts to frame society wide medical policy, are encumbered by these divisions. Health care policy is fashioned out of the characteristics of the particular cultural and moral traditions of the underlying community. However, in the West there is no single moral tradition that binds all in a common normative or evaluative understanding. Policy must be created to span a diverse set of individuals and communities. Framing health care policy involves fashioning agreements that engender a common, and usually artificial, medical, social, and moral reality. In contrast, much of Japanese traditional culture remains a strong force for directing its bioethics so that such a common reality need not be fashioned but can be found as already available.

The essays of Baruch Brody, Robert Veatch, and Tom Beauchamp advance philosophical arguments concerning the character of the moral world and its significance for medicine as if there were an uncontroversial common moral reality to which all could appeal. Such assumptions, however, cannot heal the fragmented character of moral reality. It takes little empirical research to document that moral fragmentation is significant. One need only consider the acrimonious bioethical debates

Kazumasa Hoshino (ed.), Japanese and Western Bioethics, 201–223.

concerning abortion, health care allocation and euthanasia. H. Tristram Engelhardt, Jr. and Kevin Wm. Wildes, S.J. acknowledge this fragmentation while others, such as Veatch and Beauchamp, continue to argue that a common ethical ground exists upon which a single content-full moral perspective can be erected. Europe also grapples with this splintering of moral visions. Indeed, in Europe the Enlightenment hope of disclosing a single content-full morality binding upon all human persons remains alive. This effort is reflected in European attempts to capture and legislate a singular thick secular morality. Kurt Bayertz, for example, lays out a series of arguments each of which presumes the moral authority of a single content-full moral community. It can be empirically shown, however, that a European state wide moral community does not exist. There are, for example, substantial European communities of socialists, Christian democrats, and liberals with a significant cleft between Christian and non-Christian moral and metaphysical understandings. In stark contrast to both North America and Europe, much of Japanese culture remains traditional with significant common moral understandings including commitments both to family and society. Still, as the essays in this volume demonstrate, even in Japan the strains of moral pluralism have begun to thin traditional morality. Both the East and the West are moving from traditional content-full moral communities with binding moral commitments, into the fragmentation of a post-traditional secular society.

Four points concerning these essays bear special notice. First, the North American essays reflect the post Enlightenment moral fragmentation which marks post-modernity. Even Beauchamp, who strives to unite individuals through an appeal to middle level principles, and Joy Penticuff, who invites the development of a single ethic to bind and guide nursing care, are in fact reflective of and responding to a fragmented cultural and moral reality. Each is thrust into a secular pluralistic context in which decisions regarding medical futility, scarce resource allocation, and death and dying, are made in conjunction with moral strangers. Second, the philosophical struggles of many European writers continue to reflect that Enlightenment hope for a single content-full canonical morality that will bind all humans as such. There is the expectation, at least in European bioethics, that rational argument will deliver a content-full morality to guide genetic germline research by drawing on a variety of consequentialist and deontological arguments derived from secular and religious sources. Third, I will address the arguments and reflections offered by the Japanese authors with respect to their response to the moral fragmentation within their own traditionally coherent moral culture. The Japanese authors underscore the need for adherence to a traditional culture with traditional mores and etiquette. Each ties moral norms to the context of traditional culture. A closer look, though, reveals a culture examining its own conceptual and moral presuppositions; struggling to clarify its views of reality and values, seeking clarity through rationally consistent understandings of choices which reflect deep differences in fundamental moral values. As already mentioned, fragmentation and thinning of traditional culture and values marks even Japan. Finally, I will argue that moral strangeness must be understood with reference to metaphysical and epistemological relativism. While epistemological relativism is definitive of secular bioethics, it does not necessarily involve the hopeless circumstance of metaphysical relativism.

II. WESTERN REFLECTIONS: STRUGGLING WITH PLURALISM

To establish a choice as preferable requires determining preferability regarding whom and with respect to what set of criteria. An articulated answer requires an actual determination of facts and a specification of the moral criteria employed. Insofar as one recognizes that there are a number of equally defensible but quite different moral perspectives, one loses hope of a unique authoritative secular viewpoint from which to justify particular decisions (Engelhardt, 1996a). Particular moral content is gained at the price of universality. Securing a particular ethic, such as those argued for by Penticuff, Veatch, and Beauchamp, requires specification of premises and content that will not be acknowledged as inter-subjectively morally authoritative among moral strangers.

Joy Penticuff, for example, struggles against the morally fragmented character of the contemporary world in order to secure a practical nursing ethic; an ethic in which abstract moral principles gain content through everyday understandings of what it means to the patient himself to be ill, to recover, or to move towards eventual death with a chronic terminal illness. It is argued that a nurse's moral perspective is gained at the level of everyday consciousness. Nurses deal directly with patients with a closeness and intimacy, "Nursing perspectives therefore are practical, and include all the thoughts, feelings, and actions that go into caring and being responsible for those who need nursing" (Penticuff, 1996, p. 50). Thoughts and feelings are fused with practical presence and human interaction. Deep empathetic connections with patients prompt the move from a dispassionate ethic of the primacy of individual freedom and the commodification of health care, to an ethic of a caring community, a "transformation of health care systems into caring communities – places and systems in which the good for patients can be instantiated and worked out" (p. 55). Yet, notions of the good for patients often conflict with notions of the good for providers, families, and society. Patients and providers do not operate in a vacuum in which the good of the patient is the only interest that ought to be considered. Each treatment choice allocates resources to some patients at the expense of decreasing resource availability for others. Medical choices, themselves, allocate resources away from other non medical needs.[1]

In addition, such choices require taking into account the "best interests" of patients. The difficulty is in specifying the content of "best interests." Consider, for example, two conflicting resource allocation policies each of which purports to secure the best interests of patients. The first policy offers high morbidity/low mortality health care. Resources are allocated to expensive life prolonging treatment and end of life care, including organ transplantation, kidney dialysis, and intensive care for severely deformed and anacephalic neonates. Resource availability for non emergency medical activities is limited. For instance, treatments that increase mobility, such as knee and hip replacements, are either unavailable or must be paid for out-of-pocket. The second policy offers low morbidity/high mortality health care to enhance life options, limit pain, increase mobility, and enhance freedom from illness. Treatments include knee and hip replacements, vision and hearing care, inoculation from disease, and other forms of preventative health care. This policy limits access to high cost/low yield end of life treatments, and allocates no resources to kidney dialysis or organ transplantation. As Allan Gibbard makes the point: "It may be a better prospect – rationally to be preferred, that is, on prudential groups – to enjoy what the premium will buy if one is healthy and risk needing the treatment and not being able to get it, than to live less well if healthy and get the treatment if one needs it" (1987, p. 193). Advocates of each policy argue that it implements health care that is in the patient's "best interest." Which type of policy should guide biomedical decision making? Given the many possible responses, determining best interests requires defining "best interests," specifying "for whom" and "by what criteria." Physicians, nurses, families, and patients each provide a different perspective on a multi-faceted and differentiated set of consequences, rights, liberties, and ethical solutions. Being attentive to the physical and emotional needs of a patient may assist in the clarification of some sources of the "good" of the patient. But, this will not wholly settle the question at issue because it will not justify a unique, much less necessarily correct, understanding of the "good."

Emotive caring has the benefit of reaching towards the humanity of the patient with tenderness and compassion. It struggles, however, under the burden of needing to differentiate between the patient's and the provider's perhaps very different notions of well being and humanity.[2] Also, caring may involve disengagement from medical, societal and familial reality. How are we to understand appropriately the humanity of the patient? In addition, such emotive accounts of morality collide with accounts such as that of Immanuel Kant, who argued for an objective rational morality, which is to be articulated without direct reference to human feelings and inclinations. A perfectly rational, autonomous, Kantian moral agent would necessarily act on the objective necessity of the moral imperatives. Acting on emotions or other inclinations assimilates content-full maxims that are subjectively contingent, would not be objectively universalizable for all rational agents, and would involve heteronomy, rather than free, autonomous, moral action. Thus for Kant acting on emotion or other personal inclination is of no moral worth (Kant, 1990, [1785], p. 13). Those who argue for emotive accounts regard Kant's position as removing human feeling from the realm of morality as "a disreputable, arbitrary irrelevant partner weakening the authority of reason" (Midgley, 1994, p. 152). If we are to argue, contra Kant, that our natural feelings should be embraced as important, formative elements of morality, then we must have some understanding of the rules for identifying and ranking the emotions that are to qualify. Do all emotive reactions, by all parties, point to important moral features of a situation? Which emotions should we take as authoritative and how ought we quantify them intra- and inter-personally? And if the emotive reactions are different how can one identify those that are canonical?

A similar problem presents itself in Keyserlingk's account of the moral authority of physicians. He argues: "It does appear to me slightly absurd that many of us seem quite prepared to accept a degree of moral guidance from many who have seen and shared far less of life, death, suffering, human frailty, commitment and courage than have many physicians" (1996, p. 113). Yet why would medical authority and experience necessarily entail moral authority? Does working closely with patients help one to develop appropriate morally focused emotions, canonical intuitions, or a preferable moral theory? Or, does one begin to cloud the truth with emotive valences that make facing medical (and perhaps moral) reality more difficult? The latter is at times clearly the case with families whose emotional responses of love or guilt render them unable to accept the medical reality and finality of the dying patient. And as Brody points out, this emotional situation is also experienced by many health care professionals (Brody, 1996, pp. 139–140). The difficulty is then to identify "good judgment." Even if we attempt to ground morality in the judgment of physicians the problem of identifying good judgment remains.

We confront the problem not only of defining the "good" but also of determining how the "good" ought to be realized. It is this difficulty that both Veatch and Beauchamp address. Veatch appeals to an emerging moral consensus that the patient is the best authority of his own best interests.

> Increasingly, a consensus is emerging over the underlying moral and legal principles for resolving these cases. When there is agreement that the patient is competent to make his or her own medical decisions and the treatment is proposed for the patient's own good, the autonomous choice of the individual patient to refuse medical treatment has in every single case been found to take precedence over the desire to prolong the patient's life (Veatch, 1996, p. 120; 1989; Meisel, 1992).

According to Veatch, there is a consensus forming that insofar as the patient is competent, he is the best judge of his own best interests and his decision (at least to refuse treatment) should in the end trump, as long as no significant interests of others are at stake.

There are, however, difficulties with this approach. First and foremost, one must determine who is party to the consensus as well as how and why it should bind those who do not concur with it. How in particular should one respond to conscientious objectors to the consensus? Does one use coercive force so that autonomous non-consenters accept the general consensus or should they be allowed peaceably to ignore the community's consensus?

Veatch's account of patient autonomy collides with an obligation of forbearance in the sense that the patient is in authority over his own person and ought not to be touched or interfered with without permission. Forbearance rights understood as part of the necessary grammar of morality (Engelhardt, 1996a, pp. 102ff), or as necessary side constraints (Nozick, 1975, pp. 26–35) do not have any particular content and do not depend on a moral consensus. Rather, forbearance rights are advanced as constitutive parts of the very practice of morality. They are meant to support a set of rights to be left alone that strictly limit the moral authority of physicians, health care workers and society to interfere with a patient except in those instances in which the patient consents to the treatment. By an appeal to such rights patients are able to refuse medical treatment that they find burdensome or inappropriate. Often captured under the terminology of autonomy, forbearance rights limit the moral authority of others to force non-consenters to act in conformity with general agreement. Insofar as objectors to the consensus, which Veatch advances, act peaceably with consenting collaborators the moral authority for others, including the state, to interfere is limited. Moral foundations and particular content forged through a consensus will likely only form a stable structure for social policy if it is coercively enforced upon unconsenting dissenters, thus failing to respect the fundamental differences among persons and content-full moral communities.

Beauchamp, like Veatch, attempts to establish a common content-full normative morality. He contends that at root "no difference exists in basic moral precepts between Japanese and Western morality" (1996, p. 25). There are three implicit assumptions involved in his argument: (1) that there is a single "Japanese morality," (2) that there is a single "Western morality," and (3) that "Western morality" and "Japanese morality" when compared can be shown to be morally equivalent. Beauchamp begins his analysis with a strong endorsement of moral realism: "I do not deny that this view [that of moral divergence] is valid as an expression of differences *in degree* between American and Japanese society, but I believe it has no merit as an account of differences *in kind*" (p. 30). To establish that there is a single "Japanese morality" Beauchamp cites a variety of studies showing that a large percentage of polled Japanese physicians agreed that it is important to reveal to patients their diagnosis and prognosis as well as to obtain permission before commencing treatment. The issue at stake is what to make of the physicians who acted differently. Moreover, it would be important to determine whether those who acted the same or differently did so on the basis of the same principles, much less according to the same underlying morality or moral practice. For example, Hoshino reports that "the Japanese tend not to be sensitive enough to the other's rights to autonomous decisionmaking and they override others' individual rights without realizing that they are actually behaving unethically in the American sense" (1996, p. 20). Or, as Hinohara makes the point, not informing patients about cancer originates from thoughtfulness on the part of the family; the family's feelings supersede the patient's choice as to how he should spend the rest of his life (1996, p. 153). Are middle level principles, such as autonomy and beneficence, presupposed, accepted or applied in a manner consistent with Western bioethics, or are very different moral practices grounding Japanese decision making? On Hinohara's account the physician is acting primarily under the direction of the family rather than obtaining the informed consent of the patient. This practice is not in accord with contemporary medical practice in the United States, nor does it apparently reflect similar middle level moral principles. The existence of a consensus on the part of a group of physicians does not in itself establish the existence of anything other than that particular agreement. It does not necessarily offer evidence of any fundamental moral reality, cross-cultural moral norm or "true" universal moral middle level principles. Even among the limited set of physicians polled, each may agree with the notion of informed consent because of a wide variety of different underlying moral practices, middle level principles, moral side constraints and non-moral reasons.

In addition, are we confident that moral terms have the same meaning in Japan as they do in Western countries, (presuming that moral terms are unambiguous in Western countries)? Just as it is problematic to insist on a single understanding of "good," the notion of "consent" has diverse meanings in different contexts. If a physician asks a patient for permission to begin chemotherapy, consent requires a positive statement of acceptance. In other contexts, consent is closer in meaning to assent, such as when a physician informs a patient with end-stage cancer that CPR will likely result in a flailed chest, causing more harm than good, and, therefore, that it is not a viable option. Consent requires only that the patient assent to this judgment. Consent, in this case, is more akin to "acquiescence" if a patient does not wish to be burdened with the details of his diagnosis or prognosis and does not explicitly refuse treatment. Consent can also imply only a lack of strenuous disagreement as, for example, when a patient must be manipulated and cajoled into compliance with a treatment. Consent can also be understood as the lack of rational objection. If an unconscious individual is driven to an emergency room exsanguinating from a gun shot wound, is not a documented practicing Jehovah Witness and yet a relative refuses to consent to a blood transfusion, the physician ought to understand arrival at the emergency room, together with the lack of clear and convincing evidence of a rational objection, as grounds for presuming consent to treatment. Similarly, "informed" might mean being provided that amount of information other men and women practicing medicine give to their patients; or, that amount of information reasonable men and women would need in order to accept a treatment, refuse a treatment or choose an alternative; or, that amount of information this particular patient would need in order to accept a treatment, refuse a treatment, or choose an alternative. Since standards of practice, or even of being a reasonable man or woman, can vary across contexts and communities, these three meanings of being "informed" are at best general rubrics. Given the contextual and ambiguous nature of "informed" and "consent," the exact meaning of the statement "it is important to obtain consent prior to invading a person's body," to which a Japanese physician agreed in a survey, is unclear. The usage of similar terminology does not in itself establish underlying moral realities, or the presupposition of similar middle level principles. Ambiguities in meaning and the numerous possible explanations for the existence of a consensus suggest that our ability to attain true knowledge of fundamental moral realities or of canonical middle level principles (through a consensus) is epistemologically problematic.

Finally, is there a "Western morality" that can be characterized by the same middle level principles? Since the 1970's in the United States there has been developing a consensus among physicians that they are morally required to obtain informed consent prior to treating patients. Again, what precisely does this emerging consensus reveal about this principle and similar putative middle level principles? Middle level principles appear to be either content-less or overly particular. Such principles can assist in the fashioning of agreement among those who already share similar moral understandings. Appeal to middle level principles can aid moral friends in the creation of standards of conduct, codes of ethics, and dispute settlement. However, among moral strangers there will be no agreement on moral content, or on the meaning or ranking of middle level principles. For example, consider attempting to resolve controversies by an appeal to a middle level principle of justice when the dispute involves an egalitarian and a Nozickian libertarian. Conflict resolution through appeals to consensus and to middle level principles is generally limited to discussions among moral friends. If anything, middle level principles can show in fact that there is no consensus or agreement regarding a fundamental background moral vision.

Like Veatch, Beauchamp also pushes his search towards core universal ethical principles (1996, p. 26f). Yet even in the United States, the "emerging consensus," to which both Veatch and Beauchamp refer, requires much more definition. Indeed, it is unclear how much agreement actually exists. There are, for example, fundamental disagreements among physicians about the morally appropriate manner in which to obtain informed consent. There is a lack of agreement regarding cardinal issues, such as the amount of information to be provided to patients, the appropriate mental status of the patient, the importance of family wishes, or of the evidence of a stable set of values and goals. A particularly wide and substantive consensus is unlikely to be the case. Typically, even the existence of a legally mandated standard does not thereby reflect a society-wide *moral* consensus, or even a *de facto* standard in practice. The notion that an emerging consensus reflects deeper moral content is tenuous indeed.

In part, the problem is that the very significance of informed consent is reflective of the manner in which the typical medical encounter is encumbered by moral strangeness. In Western health care, patients and physicians typically meet as moral strangers with insufficient common moral grounds for a substantive agreement regarding the other's "best interest" or "good." The *formal procedural standards* of informed consent help to ensure that patient and physician are only used in ways to which each agrees. These procedures do not, in themselves, convey any particular moral content to the decision reached or to the best interests at stake. Moreover, in other cultures the very process of informed consent may appear to the patient as harming the patient's "best interests." This has been noted even among Apache patients within the United States (Carrese and Rhodes, 1995, pp. 826–829). So too, Japanese patients may never raise the question of authorizing the physician's acts since the physician is already regarded as established in authority by a common content-full community. Moreover, the physician may be regarded as being a moral authority and able to supply the correct content for paternalistic choices. Here the Japanese appear to be in agreement with Keyserlingk. Such considerations would fundamentally recast Western concerns regarding informed consent, embedding them in the context of cultural and familial commitments.[3]

As Kevin Wm. Wildes rightly points out, while moral terms such as "sanctity of life," "respect for life," and "human dignity" have been employed to fashion a consensus and settle moral debates, such terms remain heterogenous in their meanings. This appraisal applies equally well to other appeals such as Penticuff's notion of "caring," Veatch's invocation of the patient's "best interests," and middle level principles like Beauchamp's principle "obtain consent before invading another person's body." These appeals at best capture groups of ambiguous and disparate hopes, images, feelings, values and claims (Wildes, 1996, p. 89). The terms have no fixed content-full meaning outside of particular content-full communities. For example, both Roman Catholicism and Buddhism hold human life to be sacred. Roman Catholicism draws its traditional view from God, the source of life's sanctity. Human life is sacred and holy "because life is a gift from the Holy" (p. 91). For Buddhism, however, human life is sacred because all of life is inherently sacred. As Wildes argues, outside of a particular moral context with which to provide meaning and particular content to moral terms and principles, they do not even have the status of moral non-rigid designators.

In the modern Western world, our moral terms have lost much of their traditional significance. They have become more slogans than principles. Understandings and accounts about what it means to care, to inform, to consent, to be in one's best interests, are contextual. Modern secular society with its multiple moral communities and contexts is fragmented and without common substantive moral principles with which to make authoritative society wide moral decisions. It is understandable that Penticuff, Veatch, and Beauchamp are engaged in a search for moral consensus, and for an unique content-full moral reality. However, there are good grounds for significant epistemological skepticism that a substantially wide consensus will be derived that is both morally authoritative as well as authoritatively moral. These grounds for skepticism are further augmented when comparing the divergent moral communities of the East and the West.

III. LEGISLATIVE MORAL MONISM

Endeavoring to bridge the gap among moral strangers and diverse communities, European governments have attempted to establish a single "moral culture" by legal decree. As Kurt Bayertz indicates, there is a concern to legislate a single moral culture that speaks with moral authority to all. The establishment of moral values and goals has been thrust into the political arena. Legal issues including rights to property, to freedom from injury, to unfolding freely one's personality, and concepts of human dignity, define the lines of legal debate for licit uses of medical science (p. 169). This attempt has been especially significant in Germany. As Catenhusen comments: "in Germany we are particularly sensitive with regard to the application of science and technology to human beings because of our experience with the Third Reich. Human dignity leads to the rejection of positive eugenics . . . the attempt to breed the human being" (Catenhusen, 1988, p. 40). With regard to bioethical issues, such as germ-line genetic engineering, artificial reproduction and human organ sales, European legislatures are attempting to create a content-full biopolitics that authoritatively binds all. As a consequence there is a general prohibition against genetic enhancement to increase personal performance, along with a prohibition against any modification not directly aimed at modifying generic characteristics related to a disease (Bayertz, p. 174). German legislation, focusing on ownership and uses of the human body and germ-line, provides legally protected status to a content-full body of rights and values broadly influenced by Christianity.

The problem lies in establishing the moral authority in order coercively to impose *that particular moral culture*. If the state is to draw a morally authoritative line between appropriate and inappropriate uses of the body, body parts, artificial reproduction and genetic engineering, then it must have moral authority to impose such decisions. Germ-line therapy, for example, offers a large variety of as yet untapped interventions into the human species, including treatments for genetic disease and modifications to the potential benefit of the human genome. With appropriate research, individuals may be able genetically to modify personal characteristics to maximize their, or their children's, ability to flourish. Yet, modifications will be understood by some to be beneficial morally licit "up-grades" of the human genome and by others as representing grave moral evil.[4] Without a valid demonstration that this legislation has moral authorization, from a purely secular perspective, restrictions on such therapy can only be legally, not morally, authoritative. Similarly, some forms of artificial reproduction, such as *in vitro* fertilization, will be understood by some as morally illicit, and by others as morally benign. In the absence of a successful sound rational argument or an appeal to an authority acceptable to all to show that this *particular moral culture* ought to be privileged, such legislation will be without general secular moral authority.

However, in many regards Western Europe appears committed to preserving significant elements of a moral vision which could only find a deep justification in a religious account. This European phenomenon appears to reflect a desperate attempt to recapture moral content and guidance which has been lost as European culture completes a profound transition towards becoming post-Christian. As Engelhardt argues, the European bioethical project reflects an attempt to secure, by reason and legislation, elements of the Christian tradition outside of an exclusively Christian community and without acknowledgment of the Christian Deity (Engelhardt, 1995, pp. 29–47). It is a position with great similarities to a Kantian rationalism: the kingdom of grace becomes equivalent to the kingdom of reason (Kant, 1965 [1781], A812=B840), and moral reason is given a particular content. Europe can no longer claim that it is Christendom; a Christian community with a single, content-full, coherent and all encompassing morality. Instead it contains a plurality of moral cultures. Recasting morally controversial topics, such as genetic engineering and artificial reproduction, within the legal arena offers a socio-political pragmatic resolution to controversial moral dilemmas. However, given Engelhardt's argument (1996b) this can best be understood as an attempt to secure elements of the Christian culture that fashioned Western Europe, without recourse to either religious faith or acknowledgment of the Deity. The result is a cardinal difficulty in providing a morally authoritative foundation for the particular content that is legally enforced.[5]

IV. JAPANESE REFLECTIONS: A TRADITION STRUGGLES

Japanese appeals to moral authority are often couched in terms of the content-full norms of their commitments to community and family. Many of the Japanese authors in this volume capture traditional Japanese understandings of death, dying, and health care decisions. Consider, for example, Emiko Namihira's emphasis on the necessity of traditional cultural and moral understandings for a Japanese style of cremation and bone storing rituals as these exist within orthodox Japanese funeral rites. Each storage locality, burial facility, kit for cleansing and purifying a dead body, and crematorium is integral of a traditional transition from being a living, breathing, material, bodily person to a living revered ancestor in the world of the dead. Removal of organs from the deceased body has traditionally been held to mar the identity of the person. For the Japanese, cremation is not merely the disposal of a dead body, but rather a way to change the identity of the body from living to dead. "Therefore, for Japanese people removing an organ from a dead body before such procedures have taken place represents a failure to create a new bodily identity for the dead person" (1996, p. 68). Insofar as physicians, patients and families adhere to traditional understandings of respect for the dead there is a conflict between traditional ritual and the modern medical drive for human organ availability for transplantation. In addition, the ritual impurity associated with death is being modified or lost. Traditional spirituality obliged family mourners to remain at home for 49 days after a funeral in order to avoid contaminating other people or places. Modern Japanese law has recast this ritual obligation into a legally protected leave of absence from work and school (p. 64). However, urban areas harbor families of individuals in which rituals are ignored and traditional understandings about spiritual contamination have been lost. Japanese culture is moving to a post-traditional period.

Continuing this discussion, Gen Ohi's reflections on changing attitudes about informing patients with end stage cancer of their diagnosis and prognosis indicates that the ethos of Japan is changing very slowly. Rather than becoming radically individualistic, much of Japanese society remains profoundly traditional with a community ethos that sustains mutual support with strong ties to family and community. Western principles of non-maleficence and beneficence are often recast within the Japanese context and rendered in keeping with such values. Transition towards a Western individualistic "cold" life has been resisted. The concerns to respect autonomy with which Ohi struggles are counter-traditional and in conflict with customary Japanese cultural and moral commitments. Personal autonomy is, at least *prima facie*, antagonistic to the trust inherent in a culture based upon mutual support.

The Japanese culture of mutual support with its particular content-full moral understandings embraces even expectations for health care policy concerning the harvesting of organs from deceased donors. As Kazumasa Hoshino points out, even if an individual leaves instructions authorizing organ donation upon death, in practice organs "may only be removed from the donors with the written consent of the surviving family" (p. 13). It is an ethos that paternalistically reaches into the medical context in order to protect the patient by providing family and physician support to avoid burdening the patient with too many details about his personal condition. Even if the physician is under a *prima facie* obligation of truth-telling, this obligation is discharged primarily through the family rather than directly to the patient. Close knit families decide with the physician on appropriate treatment given family desires, needs, goals and resources.

However, this familial model of decisionmaking is apparently being challenged. Arguments for the adoption of a more robust notion of informed consent do not reflect Japanese cultural heritage. Rather, they draw on Anglo-Saxon notions imported from the West that one ought not touch another without that other's direct permission.[6] Traditional Japanese medical encounters, as documented by Hoshino and Ohi, reflect an ethos that places the family in central authority and authorizes a paternalism that reaches out so as to care for the patient without invoking direct patient consent. Fumio Yamazaki's case study, which criticizes the physician and family for failing to obtain the informed consent of the patient, so that "he could have expressed his wishes at each stage of treatment and spent the last days of his life in content" (p. 133) indicates a reaction to traditional Japanese culture and values. This analysis of the immature state of informed consent in Japan invites the patient himself to be informed about his condition and to provide consent to treatment and in so doing runs counter to a cultural ethos in which the patient recognizes that the family, physician, and community will care for the patient in a morally appropriate fashion, in terms of a culture in which issues of patient informed consent do not directly arise.

The Japanese are moving, perhaps slowly, from a traditional to a post-traditional era. The authors in this volume are responding in part to a culture striving to clarify its views of morality and the values appropriately to be embraced in health care policy. The articles by the Japanese scholars reveal a culture examining its own conceptual and value presuppositions. In all of this, the diversity of traditional Japanese views is often not adequately recognized. Shigeaki Hinohara underscores the circumstance that "traditional" Japanese moral and social norms were never really homogenous. Japanese culture originated through Shintoism, although Confucianism, Taoism and Buddhism each played a major role in shaping understandings of life, copulation, purity and posterity. Each tradition shaped teachings regarding death, ritual, uncleanliness, burial and purification. When Western Christianity entered Japan, it introduced yet another set of rituals and understandings regarding life, death, and purification. More recently, with a greater exposure to a post-traditional Western world, funeral rites have become culturally thinned to such an extent that for many Japanese such rites represent only the weakest superficiality. They exist as taboos representing a mixture of only partially relevant sub-cultures and aesthetic customs of form without content. This reduction to mere form, without substance, without religious or even aesthetic significance, has lead some prominent Japanese to forbid any funeral ceremony upon their death (p. 153). While Hinohara focuses this critique on funeral customs, the force of the remarks are general for one must take into account the heterogeneity of even traditional Japanese culture. The pluralism drawn from Shintoism, Taoism, Confucianism, Buddhism, and Christianity has been further amplified by recent Western influences.

V. CONCLUSION

Where do these reflections lead us? Throughout this analysis I have been emphasizing the fragmented character of the moral world, pointing to ways in which even traditional, stable cultures are becoming increasingly fractured. Both Eastern and Western philosophers present arguments as if they were revealing through reason a common shared ethic for that society, or a consensus from which to forge a shared ethic. In addition, Western philosophers, such as Veatch and Beauchamp, argue as if there were an ethic which as a matter of fact is normative for both Western and Eastern cultures. As the Japanese reflect upon their own culture, Western philosophical conceptions and categories, such as informed consent, autonomy, beneficence, and particular notions of the good life, are being applied to the Japanese context. This is not a context in which there is a comfortable fit between Western categories and traditional Japanese commitments to family and community. The Japanese moral ethos for medical ethics becomes suspect once it is subjected to the imported categories. This confrontation with the contemporary biomedical American ethos has in areas engendered a crisis of faith. Ought Japanese bioethics reflect standard Western notions of autonomy and informed consent? Or, ought Japanese bioethics recapture robust Japanese traditional moral practices? The influx of Western bioethical reflections advanced as an element of a particular universal morality has made it appear as if the questions raised can only be answered with the rational categories and constructions of the West.

However, as I have argued, there is *de facto* and *de jure* moral heterogeneity in Western as well as Japanese culture. There is no single accepted set of fundamental or middle level moral principles. This is the case despite seeming agreement regarding a broad set of issues in bioethics. However, even though there appear to be commonly held principles, and the use of similar moral terminology, (e.g., as occurs in appeals to "autonomy" or "sanctity of life,") this does not, as we have seen, establish a single canonical meaning for such slogans. Indeed, as we have seen, there is no content-full ethic accepted by and binding on all. Western ethicists have struggled with the fragmented character of empirical moral reality for centuries. They approach and analyze the character of the moral world from a multitude of perspectives seeking rational arguments to support particular positions and claims, usually attempting to show they are universally valid. In the face of moral diversity, one Western reaction has been the attempt to preserve or establish a moral culture through force of law. This reaction appears to be a response to the profound deChristianization of Western Europe and North America inspired by the Enlightenment hope to capture, though reason alone, a content-full morality that canonically binds all humans. Yet, that hope appears, at least so far, unfulfilled. The Japanese are only beginning to recognize the implications of this drama for their own culture. Moreover, they are also beginning to experience the destabilization of their culture and its taken-for-granted expectations. As reflected in this volume, the Japanese response has been two fold. First, there have been attempts to deny heterogeneity, as represented by the reflections of Namihira and Ohi on the traditional ethos that binds individuals to their family and community in a thick moral culture. Second, others, as suggested by Hoshino, have focused on clarifying Japanese moral sentiments and collaborating both to sustain and form uniquely Japanese bioethics.

Where then does this lead our reflections concerning the state of Western and Japanese bioethics? First, it must be understood that the salience of moral difference in the West has lead to the fashioning of a Western bioethics that is procedure first and foremost, and therefore not committed to a particular moral content. Many American and European bioethical approaches have been recast so as to assist the collaborations of moral strangers. Western procedural ethics, such as the formal procedural requirements of informed consent, need not convey any particular moral content to morality. Rather they can be invoked simply to ensure that others are only used in manners to which they consent. Therefore, even if there is no universal content-full canonical moral ethic, relativism does not follow as a consequence as long as there is a procedural ethic that should bind all (see, e.g., Engelhardt, 1996a). Western approaches have been fashioned in order to accommodate to moral diversity and moral strangeness.

Secondly, and very significantly, this moral heterogeneity does not justify a normative or metaphysical relativism. There is a significant distinction between accepting moral and cultural pluralism as an epistemological circumstance of the contemporary world, even underscoring its importance for current biomedical, ethical and political debates, and embracing moral relativism. What is not warranted is to conclude from an epistemological relativism (e.g., "as a matter of epistemological fact there does not exist a single canonical content-full moral perspective which can be shown through reason alone to be known truly") to metaphysical relativism (e.g., "moral pluralism and moral relativism are true, i.e., there does not exist a single content-full moral perspective"). From the circumstance that one cannot demonstrate through reason alone that one knows truly the morality that should bind all, it does not follow that there is no such morality. An argument to establish an epistemological moral relativism requires only showing that there exist a variety of content-full communities, each of which disagrees on some, one or more, fundamental moral matters, such as the nature of the good, and the existence or ranking of middle level principles, and that one cannot select from among them the canonical morality on the basis of a sound rational argument. Proof of metaphysical relativism, however, requires a *positive argument* that *in principle* there can exist *no moral fact of the matter* in any situation – a positive argument that morality is a chimerical fiction. Even if there could not be shown to be a content-less universal procedural ethic, this would at best establish a relativity grounded in the ability to know which content should bind. It would not establish that there is no necessary content: an epistemological relativism or skepticism not a metaphysical relativism. The move from epistemological relativism to metaphysical relativism remains unjustified

Finally, within content-full communities questions of metaphysical or epistemological relativism do not arise. Or, if they do arise, there are accepted methodologies and authorities for settling such concerns. There are accepted answers to questions about appropriate treatments, touchings, and ways of dying that are generally unavailable outside of an intact tradition. Such communities provide members with a particular understanding of humanity, what it means to live properly, to die properly, to copulate appropriately and to collaborate morally with others. Within such a content-full moral community, with moral friends and consenting others, one will have enough in common to share moral premises and canons of rationality with which to forge and maintain a binding content-full bioethic. It is within such a strong culture, with consenting others, that it may be possible to forge a medical morality and sustain an ethos for health care that is truly Japanese.

NOTES

[1] See, for example, Menzel, (1992) who argues that the fundamental drive behind rationing in the medical context is that there are other things in life besides health care on which we can use our finite resources. One might expect, for example, that the fewer resources to which individuals have access the more likely they are to ration high cost, low impact care. "Thus poor people, if they can control the use of their resources (their own private resources plus any societal assisted ones to which they have access), would naturally ration care before wealthier people would" (p. 62). Given limited accessible resources, with regard to expensive, marginally beneficial treatment, individuals with lower income would probably choose differently than wealthier ones.

[2] Caring is contextual. What it means to care for someone, or to practice an ethic of care presupposes a context in which to understand how a patient wishes to show compassion for another. See for example, 'The truth about caring,' by Judith Shelly (1995, pp. 3, 46), and 'What is caring,' by Irene B. Alyn and Janet A. Conway (1995, pp. 7–12, 45), in which it is argued that the nurse's commitment to caring and compassion arise out of a particular account of Christian revelation and Christian duties. Compare this sense of caring to that for which Howard Curzer argues in 'Is care a virtue for health care professionals?' (1993, pp. 51–69). Curzer concludes that health care professionals do not have a duty to care for their patients. Rather, the notion of "care" should be deconstructed into notions of benevolence and "acting in a caring manner." These two categories again, however, require particular content-full notions of what it means to be benevolent and to "act in a caring manner" in order to be of use in determining appro-

priate codes of conduct. The particular Christian notions of Shelly, Alyn and Conway will not be authoritatively available to individuals who do not share their commitments to particular Christian faiths. In a secular context the revelation upon which they draw is not authoritatively available.

[3] On a related point within Orthodox Judaism see the *Jewish Compendium on Medical Ethics*, in Brody and Engelhardt, (1987). The notion that physicians must obtain informed consent prior to treating patients fails to take into account the content-full moral understandings of particular religions. "In the Jewish tradition, informed consent is not always essential. The principle that the physician acts with Divine license to heal may require him, as well as permit him, to do so. Just as it is his duty to heal, so it is the religious obligation of the patient to seek and receive healing. The patient who refuses a reasonable and appropriate medical regimen, whether through willfulness, despondency, invincible ignorance or irrational fear, is in violation of a Divine trust. The physician ... would thus not be subject to religious or ethical sanctions for not obtaining consent" (p. 128–129).

[4] There are a large variety of ethical issues in current debate. These include concerns with playing God (Ramsey, 1968), devaluing people (Kamm, 1994), devaluing the sacredness of human life (Mitchell, 1994), changes to the informational constraint on informed consent (Goldworth, 1995), psychological implications (Rhodes, 1995), and economic and social stakes (Pompidou, 1995).

[5] Compare this reasoning from socio-pragmatic legislation to understandings of morality with both natural law and legal positivist accounts of law and morality. Development of legislation in most times and places has been profoundly influenced by the local conventional morality and cultural ideals. For the legal positivist, however, this does not imply any necessary connection between law and morality (Hart, 1994). The criteria of legal validity of particular laws depends upon the legitimate exercise of the legislative process and not upon actual moral norms. The law may at times be influenced by morality, but it does not in itself speak to the licitness of moral norms. For natural law theory, somewhat crudely speaking, that which is law depends on the correct answer to some moral question. The locus classicus for this view is probably Plato's *Laws* in which he dismisses enactments that are not in the common interest as "no true laws" (IV 715B). However, the stronger, more typically quoted, versions arose out of Augustine and Aquinas. Augustine states "lex mihi esse non videtur, quae iust non fuerit" – "that which is not just does not seem to me to be a law" (Finnis, p. 363). Modern interpreters shorten Aquinas' stronger version of this to " Lex iniusta non est lex" – "An unjust law is not a law" (see Kretzmann, 1988). An unjust or immoral law would be so far removed from the notion of law that it cannot be understood as truly law. Notice that the German legislative response is quite removed from either of these two basic theories. The socio-pragmatic culture has influenced law to such an extent that the legislative process has become an attempt to legislate a single answer to questions regarding human genetic engineering. This single answer, based loosely on Christian concerns, is then understood as morally licit. One might argue that German legislation is just reflecting the correct moral answers to these questions. However, given the disparity among European moral traditions and Bayertz's argument that the very history of the German people renders certain (perhaps licit) answers unacceptable, it is unlikely that the legislature is in fact enacting the singularly correct moral truth.

[6] For an account of the Anglo-Saxon pagan roots of our understandings of battery

see Engelhardt, Jr. (1991), especially page 126. See also Henry Charles Lea, *Torture* for related discussions on the ancient Germanic legal immunity of free persons from torture.

BIBLIOGRAPHY

Akabayashi, A.: 1996, 'The concept of happiness in oriental thoughts and its significance in clinical medicine', in Hoshino, (ed.), *Japanese and Western Bioethics: Studies in Moral Diversity*, Kluwer Academic Publishers, Dordrecht, pp. 161–164.

Alyn, I.B. and Conway, J.A.: 1995, 'What is caring?' *Journal of Christian Nursing*, Summer, 7–12, 45.

Bayertz, K.: 1996, 'The normative status of the human genome: A European perspective', in Hoshino, (ed.), *Japanese and Western Bioethics: Studies in Moral Diversity*, Kluwer Academic Publishers, Dordrecht, pp. 167–180.

Beauchamp, T.L.: 1996, 'Comparative studies: Japan and America', in Hoshino, (ed.), *Japanese and Western Bioethics: Studies in Moral Diversity*, Kluwer Academic Publishers, Dordrecht, pp. 25–47.

Brody, B.A.: 1996, 'Medical futility: Philosophical reflections on death', in Hoshino, (ed.), *Japanese and Western Bioethics: Studies in Moral Diversity*, Kluwer Academic Publishers, Dordrecht, pp. 25–47.

Carrese, J., and Rhodes, L.: 1995, 'Western bioethics on the Navajo Reservation', *Journal of the American Medical Association* **274**, 826–829.

Catenhusen, W.M.: 1988, 'Kodifizier und der Ethik am Beispiel der Gentechnologie', in Gesellschaft Gesundheit und Forschung (eds.), *Ethik und Gentechnologie*, Main, Frankfurt.

Curzer, H.J.: 1993, 'Is care a virtue for health care professionals?' *The Journal of Medicine and Philosophy* **18**, 51–69.

Engelhardt, Jr., H.T.: 1991, *Bioethics and Secular Humanism; The Search for a Common Morality*, Trinity Press International, Philadelphia.

Engelhardt, Jr., H.T.: 1995, 'Moral content, tradition, and grace: Rethinking the possibility of a Christian bioethics', *Christian Bioethics* **1**(1), 29–47.

Engelhardt, Jr., H.T.: 1996a, *The Foundations of Bioethics*, second edition, Oxford University, New York.

Engelhardt, Jr., H.T.: 1996b, 'Moral puzzles concerning the human genome: Western taboos, intuitions, and beliefs at the end of the Christian era', in Hoshino, (ed.), *Japanese and Western Bioethics: Studies in Moral Diversity*, Kluwer Academic Publishers, Dordrecht, pp. 181–186.

Finnis, J.: 1980, *Natural Laws and Natural Rights*, The Clarendon Press, Oxford.

Gibbard, A.: 1987, 'When is the best care too expensive', in B.A. Brody and H.T. Engelhardt, Jr. (eds.), *Bioethics: Readings and Cases*, Prentice-Hall, Inc., Englewood Cliffs, N.J., pp. 192–194.

Goldworth, A.: 1995, 'Informed consent in the human genome enterprise', *Cambridge Quarterly of Health Care Ethics* **4**, 296–303.

Hart, H.L.A.: 1994, *The Concept of Law*, second edition, Clarendon Press, Oxford.

Hinohara, Shigeaki, 1996, 'Facing death the Japanese way: Customs (habit) and ethics', in Hoshino, (ed.), *Japanese and Western Bioethics: Studies in Moral Diversity*, Kluwer Academic Publishers, Dordrecht, pp. 145–154.

Hoshino, K.: 1996, 'Bioethics in the light of Japanese sentiments', in Hoshino, (ed.), *Japanese and Western Bioethics: Studies in Moral Diversity*, Kluwer Academic Publishers, Dordrecht, pp. 13–23.

Jewish Compendium on Medical Ethics, 'The best interests of the patient', in B.A. Brody and H.T. Engelhardt, Jr. (eds.), *Bioethics: Readings and Cases*, Prentice-Hall, Inc., Englewood Cliffs, N.J.

Kamm, F.: 1994, 'Moral problems in cloning embryos', *American Philosophical Association Newsletter on Philosophy and Medicine*, Spring, 91.

Kant, I.: 1965 [1781], *The Critique of Pure Reason*, N. Smith (trans.), St. Martin's Press, New York.

Kant, I.: 1990 [1785], *Foundations of the Metaphysics of Morals*, L.W. Beck (trans.), Macmillan Publishing Company, New York.

Keyserlingk, E.: 1996, 'Quality of life decisions and the hopelessly ill patient: The physician as moral agent and truth teller', in Hoshino, (ed.), *Japanese and Western Bioethics: Studies in Moral Diversity*, Kluwer Academic Publishers, Dordrecht, pp. 103–116.

Kretzmann, N.: 1988, 'Lex Iniusta Non Est Lex: – Interpreting St. Thomas Aquinas', *The American Journal of Jurisprudence*, 99–122.

Lea, H.C.: 1973 [1866], *Torture*, University of Philadelphia Press, Philadelphia.

McKenny, G.P.: 1996, 'Technology, authority, and the loss of tradition: The roots of American Bioethics in comparison with Japanese Bioethics', in Hoshino, (ed.), *Japanese and Western Bioethics: Studies in Moral Diversity*, Kluwer Academic Publishers, Dordrecht, pp. 73–87.

Meisel, A.: 1992, 'The legal consensus about forgoing life-sustaining treatment: Its status and its prospects', *Kennedy Institute of Ethics Journal* **2**, 309–345.

Menzel, P.T.: 1992, 'Some ethical costs of rationing', *Law, Medicine & Health Care* **20**(1–2), Spring–Summer, 57–66.

Midgley, M.: 1994, *The Ethical Primate; Humans, Freedom, and Morality*, Routledge, New York.

Mitchell, B.C.: 1994, 'Comment: Genetic engineering – bane or blessing?' *Ethics and Medicine* **10**(3), 50–55.

Namihira, E.: 1996, 'The characteristics of Japanese concepts and attitudes with regard to human remains', in Hoshino, (ed.), *Japanese and Western Bioethics: Studies in Moral Diversity*, Kluwer Academic Publishers, Dordrecht, pp. 61–69.

Nozick, R.: 1974, *Anarchy, State and Utopia*, Basic Books, New York.

Ohi, G.: 1996, 'Commentary on: "How the Japanese address death: custom and ethos"', in Hoshino, (ed.), *Japanese and Western Bioethics: Studies in Moral Diversity*, Kluwer Academic Publishers, Dordrecht, pp. 155–159.

Penticuff, J.H.: 1996, 'Nursing perspectives in bioethics', in Hoshino, (ed.), *Japanese and Western Bioethics: Studies in Moral Diversity*, Kluwer Academic Publishers, Dordrecht, pp. 49–60.

Pompidou, A.: 1995, 'Research on the human genome and patentability – The ethical consequences', *Journal of Medical Ethics* **21**, 69–71.

Ramsey, P.: 1968, 'The morality of abortion', in *Life or Death: Ethics and Opinions*, University of Washington Press, Seattle.
Rhodes, R.: 1995, 'Clones, harms, and rights', *Cambridge Quarterly of Health care Ethics* **4**, 285–290.
Shelly, J.A.: 1995, 'The truth about caring', *Journal of Christian Nursing*, Summer, 3, 46.
Veatch, R.M.: 1989, *Death, Dying, and the Biological Revolution*, revised edition, Yale University Press, New Haven.
Veatch, R.M.: 1996, 'Autonomy and communitarianism: The ethics of terminal care in cross-cultural perspective', in Hoshino, (ed.), *Japanese and Western Bioethics: Studies in Moral Diversity*, Kluwer Academic Publishers, Dordrecht, pp. 119–130.
Wildes, K. Wm., S.J.: 1996, 'Sanctity of life: A study in ambiguity and confusion', in Hoshino, (ed.), *Japanese and Western Bioethics: Studies in Moral Diversity*, Kluwer Academic Publishers, Dordrecht, pp. 89–101.
Yamazaki, F.: 1996, 'A thought on terminal care in Japan', in Hoshino, (ed.), *Japanese and Western Bioethics: Studies in Moral Diversity*, Kluwer Academic Publishers, Dordrecht, pp. 131–134.

NOTES ON CONTRIBUTORS

Akira Akabayashi, M.D., Assistant Professor, Department of Ethics and Anthropology, School of International Health, Faculty of Graduate Studies, The University of Tokyo, Tokyo, Japan.

Kurt Bayertz, Dr. of Philosophy, Professor of Philosophy, University of Münster, Germany.

Tom L. Beauchamp, Ph.D., Professor, Department of Philosophy, Kennedy Institute of Ethics, Georgetown University, Washington, D.C., U.S.A.

Baruch A. Brody, Ph.D., Leon Jawarsky Professor of Medical Ethics, Director, Center for Medical Ethics and Health Policy, Baylor College of Medicine; also, Professor of Philosophy, Rice University, Houston, Texas, U.S.A.

Mark J. Cherry, M.A., Managing Editor, *Christian Bioethics*, Center for Medical Ethics and Health Policy, Baylor College of Medicine, Houston, Texas, U.S.A.

H. Tristram Engelhardt, Jr., M.D., Ph.D., Professor, Departments of Medicine and Community Medicine, Center for Medical Ethics and Health Policy, Baylor College of Medicine; also, Professor of Philosophy, Rice University, Houston, Texas, U.S.A.

Ruiping Fan, B.M., M.A., Co-Managing Editor, *The Journal of Medicine and Philosophy,* Center for Medical Ethics and Health Policy, Baylor College of Medicine, Houston, Texas, U.S.A.

Shigeaki Hinohara, M.D., President, St. Luke's College of Nursing; Director, St. Luke's International Hospital; Tokyo, Japan.

Kazumasa Hoshino, M.D., D.Med.Sc., Director, International Bioethics Research Center; Professor, Institute of Religion and Culture, Kyoto's Women's University, Kyoto, Japan.

Edward W. Keyserlingk, LL.M., Ph.D., Professor, McGill University, Biomedical Ethics Unit, Montreal, Canada.

Gerald McKenny, Ph.D., Assistant Professor, Department of Religious Studies, Rice University, Houston, Texas, U.S.A.

Emiko Namihira, Ph.D., Professor of Cultural Anthropology, Kyushu Institute of Design, Fukuoka, Japan.

Gen Ohi, M.D., Professor, Department of Ethics and Anthropology, School of International Health, Faculty of Graduate Studies, The University of Tokyo, Tokyo, Japan.

Joy Hinson Penticuff, R.N., Ph.D., Associate Professor, School of Nursing, University of Texas at Austin, Austin, Texas, U.S.A.

Robert Veatch, Ph.D., Professor of Medical Ethics, Director, Kennedy Institute of Ethics, Georgetown University, Washington, D.C., U.S.A.

Kevin Wm. Wildes, S.J., Ph.D., Assistant Professor, Department of Philosophy, Georgetown University, Washington, D.C., U.S.A.

Fumio Yamazaki, M.D., Chief Doctor, St. John's Hospice, Sakuramachi General Hospital, Tokyo, Japan.

INDEX

Philosophy and Medicine

1. H. Tristram Engelhardt, Jr. and S.F. Spicker (eds.): *Evaluation and Explanation in the Biomedical Sciences.* 1975 ISBN 90-277-0553-4
2. S.F. Spicker and H. Tristram Engelhardt, Jr. (eds.): *Philosophical Dimensions of the Neuro-Medical Sciences.* 1976 ISBN 90-277-0672-7
3. S.F. Spicker and H. Tristram Engelhardt, Jr. (eds.): *Philosophical Medical Ethics.* Its Nature and Significance. 1977 ISBN 90-277-0772-3
4. H. Tristram Engelhardt, Jr. and S.F. Spicker (eds.): *Mental Health.* Philosophical Perspectives. 1978 ISBN 90-277-0828-2
5. B.A. Brody and H. Tristram Engelhardt, Jr. (eds.): *Mental Illness.* Law and Public Policy. 1980 ISBN 90-277-1057-0
6. H. Tristram Engelhardt, Jr., S.F. Spicker and B. Towers (eds.): *Clinical Judgment.* A Critical Appraisal. 1979 ISBN 90-277-0952-1
7. S.F. Spicker (ed.): *Organism, Medicine, and Metaphysics.* Essays in Honor of Hans Jonas on His 75th Birthday. 1978 ISBN 90-277-0823-1
8. E.E. Shelp (ed.): *Justice and Health Care.* 1981 ISBN 90-277-1207-7; Pb 90-277-1251-4
9. S.F. Spicker, J.M. Healey, Jr. and H. Tristram Engelhardt, Jr. (eds.): *The Law-Medicine Relation.* A Philosophical Exploration. 1981 ISBN 90-277-1217-4
10. W.B. Bondeson, H. Tristram Engelhardt, Jr., S.F. Spicker and J.M. White, Jr. (eds.): *New Knowledge in the Biomedical Sciences.* Some Moral Implications of Its Acquisition, Possession, and Use. 1982 ISBN 90-277-1319-7
11. E.E. Shelp (ed.): *Beneficence and Health Care.* 1982 ISBN 90-277-1377-4
12. G.J. Agich (ed.): *Responsibility in Health Care.* 1982 ISBN 90-277-1417-7
13. W.B. Bondeson, H. Tristram Engelhardt, Jr., S.F. Spicker and D.H. Winship: *Abortion and the Status of the Fetus.* 2nd printing, 1984 ISBN 90-277-1493-2
14. E.E. Shelp (ed.): *The Clinical Encounter.* The Moral Fabric of the Patient-Physician Relationship. 1983 ISBN 90-277-1593-9
15. L. Kopelman and J.C. Moskop (eds.): *Ethics and Mental Retardation.* 1984 ISBN 90-277-1630-7
16. L. Nordenfelt and B.I.B. Lindahl (eds.): *Health, Disease, and Causal Explanations in Medicine.* 1984 ISBN 90-277-1660-9
17. E.E. Shelp (ed.): *Virtue and Medicine.* Explorations in the Character of Medicine. 1985 ISBN 90-277-1808-3
18. P. Carrick: *Medical Ethics in Antiquity.* Philosophical Perspectives on Abortion and Euthanasia. 1985 ISBN 90-277-1825-3; Pb 90-277-1915-2
19. J.C. Moskop and L. Kopelman (eds.): *Ethics and Critical Care Medicine.* 1985 ISBN 90-277-1820-2
20. E.E. Shelp (ed.): *Theology and Bioethics.* Exploring the Foundations and Frontiers. 1985 ISBN 90-277-1857-1
21. G.J. Agich and C.E. Begley (eds.): *The Price of Health.* 1986 ISBN 90-277-2285-4
22. E.E. Shelp (ed.): *Sexuality and Medicine.* Vol. I: Conceptual Roots. 1987 ISBN 90-277-2290-0; Pb 90-277-2386-9

23. E.E. Shelp (ed.): *Sexuality and Medicine.* Vol. II: Ethical Viewpoints in Transition. 1987 ISBN 1-55608-013-1; Pb 1-55608-016-6
24. R.C. McMillan, H. Tristram Engelhardt, Jr., and S.F. Spicker (eds.): *Euthanasia and the Newborn.* Conflicts Regarding Saving Lives. 1987 ISBN 90-277-2299-4; Pb 1-55608-039-5
25. S.F. Spicker, S.R. Ingman and I.R. Lawson (eds.): *Ethical Dimensions of Geriatric Care.* Value Conflicts for the 21th Century. 1987 ISBN 1-55608-027-1
26. L. Nordenfelt: *On the Nature of Health.* An Action-Theoretic Approach. 2nd, rev. ed. 1995 ISBN 0-7923-3369-1; Pb 0-7923-3470-1
27. S.F. Spicker, W.B. Bondeson and H. Tristram Engelhardt, Jr. (eds.): *The Contraceptive Ethos.* Reproductive Rights and Responsibilities. 1987 ISBN 1-55608-035-2
28. S.F. Spicker, I. Alon, A. de Vries and H. Tristram Engelhardt, Jr. (eds.): *The Use of Human Beings in Research.* With Special Reference to Clinical Trials. 1988 ISBN 1-55608-043-3
29. N.M.P. King, L.R. Churchill and A.W. Cross (eds.): *The Physician as Captain of the Ship.* A Critical Reappraisal. 1988 ISBN 1-55608-044-1
30. H.-M. Sass and R.U. Massey (eds.): *Health Care Systems.* Moral Conflicts in European and American Public Policy. 1988 ISBN 1-55608-045-X
31. R.M. Zaner (ed.): *Death: Beyond Whole-Brain Criteria.* 1988 ISBN 1-55608-053-0
32. B.A. Brody (ed.): *Moral Theory and Moral Judgments in Medical Ethics.* 1988 ISBN 1-55608-060-3
33. L.M. Kopelman and J.C. Moskop (eds.): *Children and Health Care.* Moral and Social Issues. 1989 ISBN 1-55608-078-6
34. E.D. Pellegrino, J.P. Langan and J. Collins Harvey (eds.): *Catholic Perspectives on Medical Morals.* Foundational Issues. 1989 ISBN 1-55608-083-2
35. B.A. Brody (ed.): *Suicide and Euthanasia.* Historical and Contemporary Themes. 1989 ISBN 0-7923-0106-4
36. H.A.M.J. ten Have, G.K. Kimsma and S.F. Spicker (eds.): *The Growth of Medical Knowledge.* 1990 ISBN 0-7923-0736-4
37. I. Löwy (ed.): *The Polish School of Philosophy of Medicine.* From Tytus Chałubiński (1820–1889) to Ludwik Fleck (1896–1961). 1990 ISBN 0-7923-0958-8
38. T.J. Bole III and W.B. Bondeson: *Rights to Health Care.* 1991 ISBN 0-7923-1137-X
39. M.A.G. Cutter and E.E. Shelp (eds.): *Competency.* A Study of Informal Competency Determinations in Primary Care. 1991 ISBN 0-7923-1304-6
40. J.L. Peset and D. Gracia (eds.): *The Ethics of Diagnosis.* 1992 ISBN 0-7923-1544-8

Philosophy and Medicine

41. K.W. Wildes, S.J., F. Abel, S.J. and J.C. Harvey (eds.): *Birth, Suffering, and Death.* Catholic Perspectives at the Edges of Life. 1992 [CSiB-1]
    ISBN 0-7923-1547-2; Pb 0-7923-2545-1
42. S.K. Toombs: *The Meaning of Illness.* A Phenomenological Account of the Different Perspectives of Physician and Patient. 1992
    ISBN 0-7923-1570-7; Pb 0-7923-2443-9
43. D. Leder (ed.): *The Body in Medical Thought and Practice.* 1992
    ISBN 0-7923-1657-6
44. C. Delkeskamp-Hayes and M.A.G. Cutter (eds.): *Science, Technology, and the Art of Medicine.* European-American Dialogues. 1993 ISBN 0-7923-1869-2
45. R. Baker, D. Porter and R. Porter (eds.): *The Codification of Medical Morality.* Historical and Philosophical Studies of the Formalization of Western Medical Morality in the 18th and 19th Centuries, Volume One: Medical Ethics and Etiquette in the 18th Century. 1993 ISBN 0-7923-1921-4
46. K. Bayertz (ed.): *The Concept of Moral Consensus.* The Case of Technological Interventions in Human Reproduction. 1994 ISBN 0-7923-2615-6
47. L. Nordenfelt (ed.): *Concepts and Measurement of Quality of Life in Health Care.* 1994 [ESiP-1] ISBN 0-7923-2824-8
48. R. Baker and M.A. Strosberg (eds.) with the assistance of J. Bynum: *Legislating Medical Ethics.* A Study of the New York State Do-Not-Resuscitate Law. 1995 ISBN 0-7923-2995-3
49. R. Baker (ed.): *The Codification of Medical Morality.* Historical and Philosophical Studies of the Formalization of Western Morality in the 18th and 19th Centuries, Volume Two: Anglo-American Medical Ethics and Medical Jurisprudence in the 19th Century. 1995 ISBN 0-7923-3528-7; Pb 0-7923-3529-5
50. R.A. Carson and C.R. Burns (eds.): *The Philosophy of Medicine and Bioethics.* Retrospective and Critical Appraisal. (forthcoming) ISBN 0-7923-3545-7
51. K.W. Wildes, S.J. (ed.): *Critical Choices and Critical Care.* Catholic Perspectives on Allocating Resources in Intensive Care Medicine. 1995 [CSiB-2]
    ISBN 0-7923-3382-9
52. K. Bayertz (ed.): *Sanctity of Life and Human Dignity.* 1996
    ISBN 0-7923-3739-5
53. Kevin Wm. Wildes, S.J. (ed.): *Infertility: A Crossroad of Faith, Medicine, and Technology.* 1996 ISBN 0-7923-4061-2
54. Kazumasa Hoshino (ed.): *Japanese and Western Bioethics.* Studies in Moral Diversity. 1996 ISBN 0-7923-4112-0

KLUWER ACADEMIC PUBLISHERS – DORDRECHT / BOSTON / LONDON